NEW PLANE AND SPHERICAL TRIGONOMETRY

NEW PLANE AND SPHERICAL TRIGONOMETRY

WEBSTER WELLS

GYAN

Published by

Gyan Publishing House
5, Ansari Road
Daryaganj, New Delhi-110002
Phone: 011-47034999, 9811692060
E-mail: books@gyanbooks.com

Distribution Network
gyanbooks.com
India, USA, Canada, UK, Australia

ISBN: 978-93-7414-668-2 (PB)

First Published, 1896

2nd Impression 2026

Printed at: Gyan Press, Delhi.

NEW PLANE AND SPHERICAL TRIGONOMETRY
Author: WEBSTER WELLS

NEW

PLANE AND SPHERICAL

TRIGONOMETRY

BY

WEBSTER WELLS, S.B.

PROFESSOR OF MATHEMATICS IN THE MASSACHUSETTS
INSTITUTE OF TECHNOLOGY

LEACH, SHEWELL, AND SANBORN

BOSTON NEW YORK CHICAGO

PREFACE.

In revising his Plane and Spherical Trigonometry, the author has effected many important improvements. The attention of teachers is specially invited to the following features of the new work:

1. The proofs of the functions of 0°, 90°, 180°, and 270°; §§ 22 to 25.

2. The proofs of the functions of 120°, 135°, etc.; § 27.

3. The method of finding the values of the remaining functions of an angle when the value of any one is given; § 28.

4. The proofs of the functions of $(-A)$, and $(90° + A)$, in terms of those of A; §§ 29, 30.

5. The method of solution in the examples of §§ 34 and 35.

6. The general demonstration of the formulæ for $\sin(x + y)$ and $\cos(x + y)$; § 42.

7. The discussion of the line values of the functions, and their application in tracing the changes in the six principal functions of an angle as the angle increases from 0° to 360°; §§ 60, 61.

8. The discussion of trigonometric equations in § 62.

9. The solution of right triangles by Natural Functions; see Ex. 1, page 54.

10. The discussion of the ambiguous case in the solution of oblique triangles; §§ 117 to 120.

11. The proof of the formulæ for the values of x in the cubic equation $x^3 - ax - b = 0$; § 126.

12. The *geometrical* proof of the important theorems of § 133.

13. The demonstration of the formulæ for right spherical triangles *before* those for oblique spherical triangles; see Chapters XI. and XII.

14. The reduction of the number of cases in the complete demonstration of the fundamental theorems for spherical right triangles, to three, by application of the theorems of § 133; see § 136.

15. The solution of Quadrantal and Isosceles Spherical triangles; §§ 149, 150.

16. The discussion of the ambiguous cases in the solution of oblique spherical triangles; §§ 165, 166; especially the rules given on pages 108 and 111 for determining the number of solutions.

At the end of Chapter XII. will be found a collection of formulæ in form for convenient reference.

The revised work contains a much greater number of examples than the old; they have been selected with great care, and are with few exceptions new.

The results have been worked out by aid of the author's new Six Place Logarithmic Tables, which contain also a Table of Natural Functions, and an Auxiliary Table for Small Angles. The Trigonometry can be obtained either with or without the Tables.

WEBSTER WELLS.

MASSACHUSETTS INSTITUTE OF TECHNOLOGY, 1896.

CONTENTS.

PLANE TRIGONOMETRY.

SPHERICAL TRIGONOMETRY.

PLANE TRIGONOMETRY.

I. TRIGONOMETRIC FUNCTIONS OF ACUTE ANGLES.

1. Trigonometry treats of the properties and measurement of angles and triangles.

In *Plane Trigonometry* we consider *plane* figures only.

2. Definitions of the Trigonometric Functions of Acute Angles.

Let BAC be any acute angle.

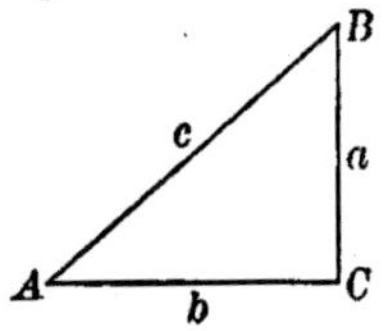

From any point in either side, as B, draw a perpendicular to the other side, forming the right triangle ABC.

We then have the following definitions, applicable to either of the acute angles A or B:

In any right triangle,

The **sine** *of either acute angle is the ratio of the opposite side to the hypotenuse.*

The **cosine** *is the ratio of the adjacent side to the hypotenuse.*

The **tangent** *is the ratio of the opposite side to the adjacent side.*

The **cotangent** *is the ratio of the adjacent side to the opposite side.*

The **secant** *is the ratio of the hypotenuse to the adjacent side.*

The **cosecant** *is the ratio of the hypotenuse to the opposite side.*

We also have the following definitions:

The **versed sine** *of an angle is* 1 *minus the cosine of the angle.*

The **coversed sine** *is* 1 *minus the sine.*

The eight ratios defined above are called the *Trigonometric Functions* of the angle.

Representing the sides BC, CA, and AB by a, b, and c, respectively, and employing the usual abbreviations, we have:

$$\sin A = \frac{a}{c}. \quad \tan A = \frac{a}{b}. \quad \sec A = \frac{c}{b}. \quad \text{vers } A = 1 - \frac{b}{c}.$$

$$\cos A = \frac{b}{c}. \quad \cot A = \frac{b}{a}. \quad \csc A = \frac{c}{a}. \quad \text{covers } A = 1 - \frac{a}{c}.$$

$$\sin B = \frac{b}{c}. \quad \tan B = \frac{b}{a}. \quad \sec B = \frac{c}{a}. \quad \text{vers } B = 1 - \frac{a}{c}.$$

$$\cos B = \frac{a}{c}. \quad \cot B = \frac{a}{b}. \quad \csc B = \frac{c}{b}. \quad \text{covers } B = 1 - \frac{b}{c}.$$

3. It is important to observe that the values of the trigonometric functions depend solely on the magnitude of the angle, and are entirely independent of the lengths of the sides of the right triangle which contains it.

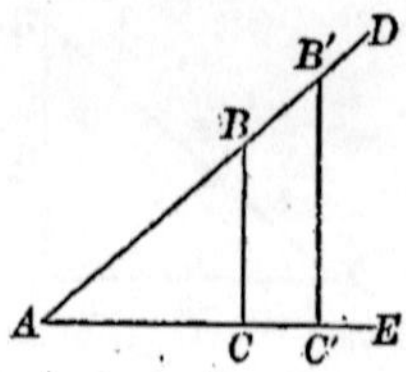

For let B and B' be any two points in the side AD of the angle DAE, and draw BC and $B'C'$ perpendicular to AE.

Then by the definition of § 2, we have

$$\sin A = \frac{BC}{AB}, \text{ and } \sin A = \frac{B'C'}{AB'}.$$

But since the right triangles ABC and $AB'C'$ are similar, their homologous sides are proportional; whence,

$$\frac{BC}{AB} = \frac{B'C'}{AB'}.$$

Thus the two values obtained for $\sin A$ are equal.

4. We have from § 2,

$$\sin A = \frac{a}{c}, \; \cos A = \frac{b}{c}, \; \sin B = \frac{b}{c}, \text{ and } \cos B = \frac{a}{c}.$$

Whence, $a = c \sin A = c \cos B$, and $b = c \sin B = c \cos A$.

That is, *in any right triangle, either side about the right angle is equal to the hypotenuse multiplied by the sine of the opposite angle, or by the cosine of the adjacent angle.*

Again, $\tan A = \frac{a}{b}$, $\cot A = \frac{b}{a}$, $\tan B = \frac{b}{a}$, and $\cot B = \frac{a}{b}$.

Whence, $a = b \tan A = b \cot B$, and $b = a \tan B = a \cot A$.

That is, *in any right triangle, either side about the right angle is equal to the tangent of the opposite angle, or the cotangent of the adjacent angle, multiplied by the other side.*

5. We have from § 2,

$$\sin A = \frac{a}{c} = \cos B. \qquad \sec A = \frac{c}{b} = \csc B.$$

$$\tan A = \frac{a}{b} = \cot B. \qquad \text{vers } A = 1 - \frac{b}{c} = \text{covers } B.$$

As B is the complement of A, these results may be stated as follows:

The sine, tangent, secant, and versed sine of any acute angle are respectively the cosine, cotangent, cosecant, and coversed sine of the complement of the angle.

6. To Find the Values of the Other Seven Functions of an Acute Angle, when the Value of Any One is Given.

1. Given $\csc A = 3$; find the values of the remaining functions of A.

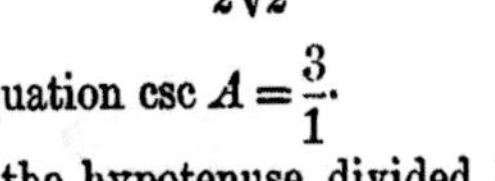

We may write the equation $\csc A = \frac{3}{1}$.

Since the cosecant is the hypotenuse divided by the opposite side, we may regard A as one of the acute angles of the right triangle ABC, in which the hypotenuse $AB = 3$, and the opposite side $BC = 1$.

Whence by Geometry, $AC = \sqrt{\overline{AB}^2 - \overline{BC}^2} = \sqrt{9-1} = \sqrt{8} = 2\sqrt{2}$.

Then by the definitions of § 2,

$$\sin A = \frac{1}{3}. \qquad \tan A = \frac{1}{2\sqrt{2}}. \qquad \sec A = \frac{3}{2\sqrt{2}}. \qquad \text{vers } A = 1 - \frac{2\sqrt{2}}{3}.$$

$$\cos A = \frac{2\sqrt{2}}{3}. \qquad \cot A = 2\sqrt{2}. \qquad \text{covers } A = 1 - \frac{1}{3} = \frac{2}{3}.$$

2. Given vers $A=\frac{2}{5}$; find the value of cot A.

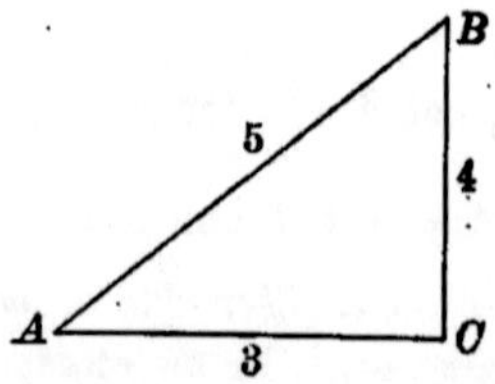

Since vers $A=1-\cos A$, we have $\cos A=1-\text{vers } A=1-\frac{2}{5}=\frac{3}{5}$.

Then, in the right triangle ABC, we take the adjacent side $AC=3$, and the hypotenuse $AB=5$.

Whence, $BC=\sqrt{\overline{AB}^2-\overline{AC}^2}=\sqrt{25-9}=\sqrt{16}=4.$

Then by definition, $\cot A=\frac{3}{4}$.

EXAMPLES.

In each of the following, find the values of the remaining functions:

3. $\sin A=\frac{3}{5}$.
5. $\cot A=\frac{7}{24}$.
7. $\cos A=\frac{3\sqrt{3}}{14}$.
9. $\sec A=x$.
4. $\text{vers } A=\frac{8}{13}$.
6. $\csc A=7$.
8. $\text{covers } A=\frac{2}{17}$.
10. $\tan A=\frac{b}{a}$.

11. Given $\cot A=\frac{3}{2}$; find $\sin A$.
14. Given $\cos A=\frac{21}{29}$; find $\csc A$.
12. Given $\csc A=\frac{41}{40}$; find $\cos A$.
15. Given $\tan A=\frac{4\sqrt{2}}{7}$; find $\sec A$.
13. Given $\sec A=5$; find $\cot A$.
16. Given $\sin A=\frac{y}{x}$; find $\tan A$.

7. Functions of 45°.

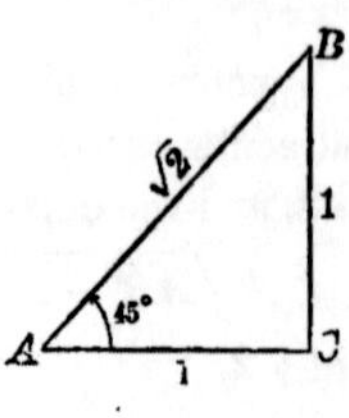

Let ABC be an isosceles right triangle, AC and BC being each equal to 1.

Then $\angle A=45°$, and $AB=\sqrt{\overline{AC}^2+\overline{BC}^2}=\sqrt{1+1}=\sqrt{2}$.

Whence by definition,

$$\sin 45^\circ = \frac{1}{\sqrt{2}} = \frac{1}{2}\sqrt{2}. \qquad \sec 45^\circ = \sqrt{2}.$$

$$\csc 45^\circ = \sqrt{2}.$$

$$\cos 45^\circ = \frac{1}{\sqrt{2}} = \frac{1}{2}\sqrt{2}. \qquad \operatorname{vers} 45^\circ = 1 - \frac{1}{2}\sqrt{2} = \frac{2-\sqrt{2}}{2}.$$

$$\tan 45^\circ = 1.$$

$$\operatorname{covers} 45^\circ = 1 - \frac{1}{2}\sqrt{2} = \frac{2-\sqrt{2}}{2}.$$

$$\cot 45^\circ = 1.$$

8. Functions of 30° and 60°.

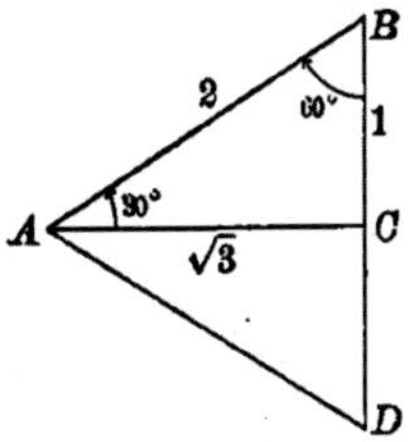

Let ABD be an equilateral triangle having each side equal to 2.
Draw AC perpendicular to BD.
Then by Geometry, $BC = \frac{1}{2} BD = 1$, and $\angle BAC = \frac{1}{2}\angle BAD = 30^\circ$.
Also, $AC = \sqrt{\overline{AB}^2 - \overline{BC}^2} = \sqrt{4-1} = \sqrt{3}$.
Then by definition, in the right triangle ABC,

$$\sin 30^\circ = \frac{1}{2} = \cos 60^\circ. \qquad \sec 30^\circ = \frac{2}{\sqrt{3}} = \frac{2}{3}\sqrt{3} = \csc 60^\circ.$$

$$\cos 30^\circ = \frac{\sqrt{3}}{2} = \sin 60^\circ. \qquad \csc 30^\circ = 2 = \sec 60^\circ.$$

$$\operatorname{vers} 30^\circ = 1 - \frac{\sqrt{3}}{2} = \operatorname{covers} 60^\circ.$$

$$\tan 30^\circ = \frac{1}{\sqrt{3}} = \frac{1}{3}\sqrt{3} = \cot 60^\circ.$$

$$\operatorname{covers} 30^\circ = 1 - \frac{1}{2} = \frac{1}{2} = \operatorname{vers} 60^\circ.$$

$$\cot 30^\circ = \sqrt{3} = \tan 60^\circ.$$

II. TRIGONOMETRIC FUNCTIONS OF ANGLES IN GENERAL.

9. In Geometry, we are, as a rule, concerned with angles which are less than two right angles; but in Trigonometry it is convenient to consider them as unrestricted in magnitude.

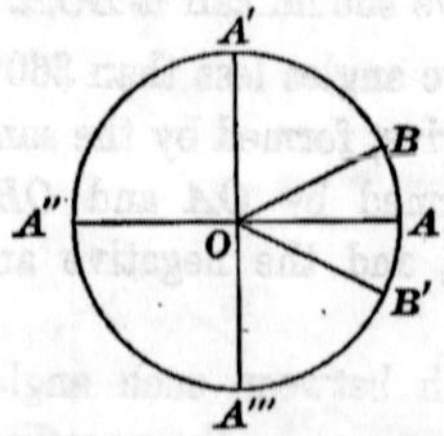

Let AA'' and $A'A'''$ be a pair of perpendicular diameters of the circle AA''.

Suppose a radius OB to start from the position OA, and revolve about the point O as a pivot, in a direction contrary to the motion of the hands of a clock.

When OB coincides with OA', it has generated an angle of 90°; when it coincides with OA'', of 180°; with OA''', of 270°; with OA, its first position, of 360°; with OA' again, of 450°; and so on.

We thus see that a significance may be attached to a positive angle of any number of degrees.

10. The interpretation of an angle as measuring the amount of rotation of a moving radius, enables us to distinguish between positive and negative angles.

Thus, if a positive angle indicates revolution from the position OA in a direction *contrary* to the motion of the hands of a clock, a negative angle may be taken as indicating revolution from the position OA in *the same* direction as the motion of the hands of a clock.

Thus, if the radius OB' starts from the position OA, and revolves about the point O as a pivot in the same direction as the motion of the hands of a clock, when it coincides with OA''' it has generated an angle of $-90°$; when it coincides with OA'', of $-180°$; with OA', of $-270°$; and so on.

We may then conceive of *negative* angles of any number of degrees.

It is immaterial which direction we consider the positive direction of rotation; but having at the outset adopted a certain direction as positive, our subsequent operations must be in accordance.

11. The fixed line OA from which the rotation is supposed to commence, is called the *initial line*, and the rotating radius in its final position is called the *terminal line.*

12. In designating an angle, we shall always write first the letter at the extremity of the initial line.

Thus, in designating the angle formed by the lines OA and OB, if we regard OA as the initial line, we should call it AOB; and if we regard OB as the initial line, we should call it BOA.

There are always two angles less than 360° in absolute value, one positive and the other negative, formed by the same initial and terminal line.

Thus, there are formed by OA and OB' the positive angle AOB' between 270° and 360°, and the negative angle AOB' between 0° and $-90°$.

We shall distinguish between such angles by referring to them as "the positive angle AOB'" and "the negative angle AOB'," respectively.

13. It is evident that the terminal lines of any two angles which differ by a multiple of 360° are coincident.

Thus, the angles 30°, 390°, $-330°$, etc., have the same terminal line.

14. Rectangular Co-ordinates.

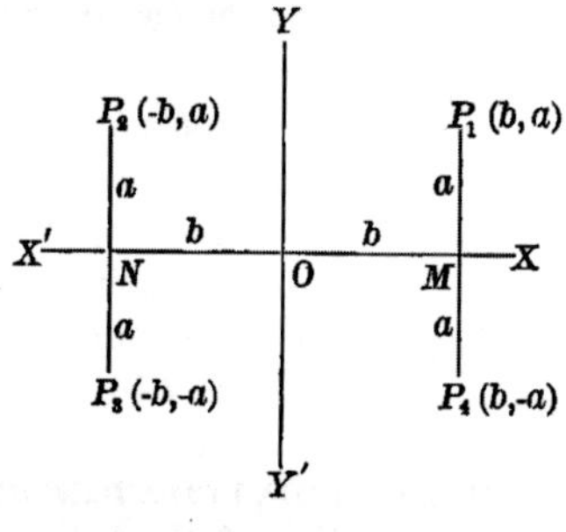

Let P_1 be any point in the plane of the lines XX' and YY' intersecting at right angles at O, and draw P_1M perpendicular to XX'.

Then OM and P_1M are called the *rectangular co-ordinates* of P_1; OM is called the *abscissa*, and P_1M the *ordinate.*

The lines of reference, XX' and YY', are called the *axis of* X and the *axis of* Y, respectively, and O is called the *origin.*

It is customary to express the fact that the abscissa of a point is b, and its ordinate a, by saying that for the point in question $x = b$ and $y = a$; or, more concisely, we may refer to the point as "the point (b, a)," where the first term in the parenthesis is understood to be the abscissa, and the second term the ordinate.

15. If, in the figure of § 14, $OM = ON = b$, and P_1P_4 and P_2P_3 are drawn perpendicular to XX' so that $P_1M = P_2N = P_3N = P_4M = a$, the points P_1, P_2, P_3, and P_4 will have the same co-ordinates, (b, a).

To avoid this ambiguity, abscissas measured to the *right* of O are considered *positive*, and to the *left*, *negative*; and ordinates measured *above* XX' are considered *positive*, and *below*, *negative*.

Then the co-ordinates of the points will be as follows:

$$P_1, (b, a);\ P_2, (-b, a);\ P_3, (-b, -a);\ P_4, (b, -a).$$

16. If a point lies upon XX', its ordinate is zero; and if it lies upon YY', its abscissa is zero.

17. General Definitions of the Functions.

We will now give general definitions of the trigonometric functions, applicable to any angle whatever.

Take the initial line of the angle as the positive direction of the axis of X, the vertex being the origin.

From any point in the terminal line, drop a perpendicular to the axis of X.

Find the co-ordinates of this point; then,

The **sine** *of the angle is the ratio of the* **ordinate** *of the point to its* **distance** *from the origin.*

The **cosine** *is the ratio of the* **abscissa** *to the* **distance.**
The **tangent** *is the ratio of the* **ordinate** *to the* **abscissa.**
The **cotangent** *is the ratio of the* **abscissa** *to the* **ordinate.**
The **secant** *is the ratio of the* **distance** *to the* **abscissa.**
The **cosecant** *is the ratio of the* **distance** *to the* **ordinate.**

18. We will now apply the definitions of § 17 to finding the functions of the angles XOP_2, XOP_3, and XOP_4 in the following figures:

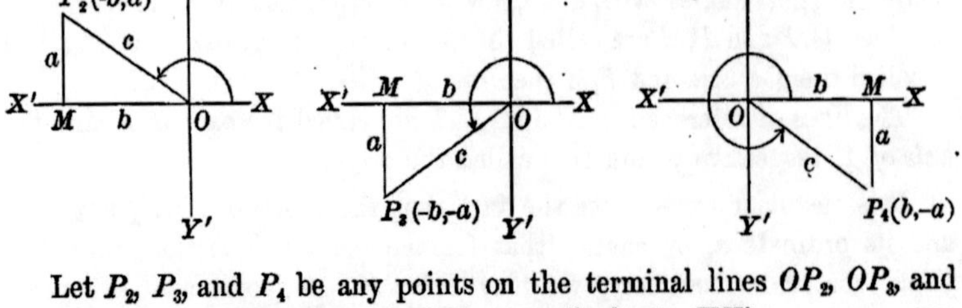

Let P_2, P_3, and P_4 be any points on the terminal lines OP_2, OP_3, and OP_4, and draw P_2M, P_3M, and P_4M perpendicular to XX'.

Let $P_2M = P_3M = P_4M = a$, $OM = b$, and $OP_2 = OP_3 = OP_4 = c$.

Then the co-ordinates of P_2 are $(-b, a)$; of P_3, $(-b, -a)$; of P_4, $(b, -a)$.

Whence by definition,

$\sin XOP_2 = \frac{a}{c}$. $\quad \sin XOP_3 = \frac{-a}{c} = -\frac{a}{c}$. $\quad \sin XOP_4 = \frac{-a}{c} = -\frac{a}{c}$.

$\cos XOP_2 = \frac{-b}{c} = -\frac{b}{c}$. $\quad \cos XOP_3 = \frac{-b}{c} = -\frac{b}{c}$. $\quad \cos XOP_4 = \frac{b}{c}$.

$\tan XOP_2 = \frac{a}{-b} = -\frac{a}{b}$. $\quad \tan XOP_3 = \frac{-a}{-b} = \frac{a}{b}$. $\quad \tan XOP_4 = \frac{-a}{b} = -\frac{a}{b}$.

$\cot XOP_2 = \frac{-b}{a} = -\frac{b}{a}$. $\quad \cot XOP_3 = \frac{-b}{-a} = \frac{b}{a}$. $\quad \cot XOP_4 = \frac{b}{-a} = -\frac{b}{a}$.

$\sec XOP_2 = \frac{c}{-b} = -\frac{c}{b}$. $\quad \sec XOP_3 = \frac{c}{-b} = -\frac{c}{b}$. $\quad \sec XOP_4 = \frac{c}{b}$.

$\csc XOP_2 = \frac{c}{a}$. $\quad \csc XOP_3 = \frac{c}{-a} = -\frac{c}{a}$. $\quad \csc XOP_4 = \frac{c}{-a} = -\frac{c}{a}$.

Note 1. The definitions of § 17 are seen to include those of § 2.

The definitions of the versed sine and coversed sine, given in § 2, are sufficiently general to apply to any angle whatever.

Note 2. In all the figures of the present chapter, the small letters will be understood as denoting the *lengths* of the lines to which they are attached, without regard to their algebraic sign.

19. If the initial line of an angle coincides with OX, and its terminal line lies between OX and OY, the angle is said to be in the *first quadrant;* if the terminal line lies between OY and OX', the angle is said to be in the *second quadrant;* if between OX' and OY', in the *third quadrant;* if between OY' and OX, in the *fourth quadrant.*

Thus, any positive angle between 0° and 90°, or 360° and 450°, or any negative angle between −270° and −360°, is in the first quadrant; any positive angle between 90° and 180°, or 450° and 540°, or any negative angle between −180° and −270°, is in the second quadrant.

20. It follows from the definitions of § 17 that, *for any angle in the first quadrant, all the functions are positive.*

It is also evident by inspection of the results of § 18 that:

In the second quadrant, the sine and cosecant are positive, and the cosine, tangent, cotangent, and secant are negative.

In the third quadrant, the tangent and cotangent are positive, and the sine, cosine, secant, and cosecant are negative.

In the fourth quadrant, the cosine and secant are positive, and the sine, tangent, cotangent, and cosecant are negative.

It is usual to express the above in tabular form, as follows:

Functions.	*First Quad.*	*Second Quad.*	*Third Quad.*	*Fourth Quad.*
Sine and cosecant	+	+	−	−
Cosine and secant	+	−	−	+
Tangent and cotangent	+	−	+	−

21. Since the terminal lines of any two angles which differ by a multiple of 360° are coincident (§ 13), it is evident that the trigonometric functions of two such angles are identical.

Thus, the functions of 50°, 410°, 770°, −310°, etc., are identical.

22. Functions of 0° and 360°.

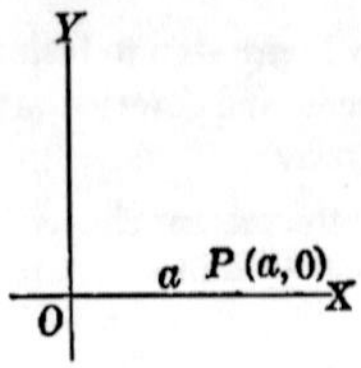

The terminal line of 0° coincides with the initial line OX.

Let P be a point on OX such that $OP = a$.

Then by § 16, the co-ordinates of P are $(a, 0)$.

Whence by definition,

$$\sin 0° = \frac{0}{a} = 0. \qquad \tan 0° = \frac{0}{a} = 0. \qquad \sec 0° = \frac{a}{a} = 1.$$

$$\cos 0° = \frac{a}{a} = 1. \qquad \cot 0° = \frac{a}{0} = \infty. \qquad \csc 0° = \frac{a}{0} = \infty.$$

By § 21, the functions of 360° are the same as those of 0°.

23. Functions of 90°.

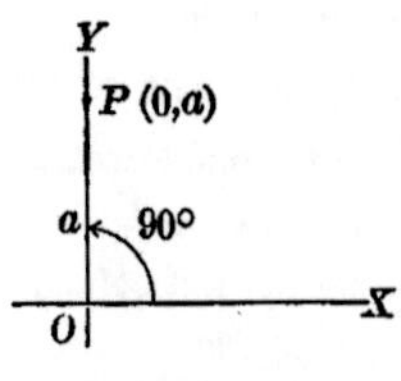

Let P be a point on OY such that $OP = a$.

Then the co-ordinates of P are $(0, a)$.

Whence by definition,

$\sin 90° = \frac{a}{a} = 1.$ $\qquad \tan 90° = \frac{a}{0} = \infty.$ $\qquad \sec 90° = \frac{a}{0} = \infty.$

$\cos 90° = \frac{0}{a} = 0.$ $\qquad \cot 90° = \frac{0}{a} = 0.$ $\qquad \csc 90° = \frac{a}{a} = 1.$

24. Functions of 180°.

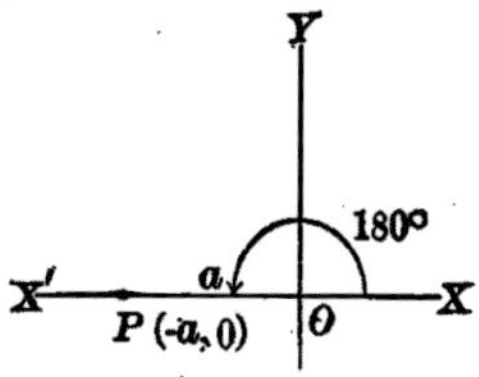

Let P be a point on OX' such that $OP = a$.
Then the co-ordinates of P are $(-a, 0)$.
Whence by definition,

$\sin 180° = \frac{0}{a} = 0.$ $\qquad \tan 180° = \frac{0}{-a} = 0.$ $\qquad \sec 180° = \frac{a}{-a} = -1.$

$\cos 180° = \frac{-a}{a} = -1.$ $\qquad \cot 180° = \frac{-a}{0} = \infty.$ $\qquad \csc 180° = \frac{a}{0} = \infty.$

25. Functions of 270°.

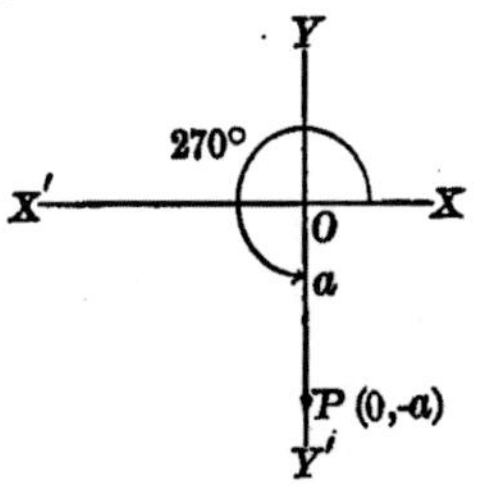

Let P be a point on OY' such that $OP = a$.
Then the co-ordinates of P are $(0, -a)$.
Whence by definition,

$\sin 270° = \frac{-a}{a} = -1.$ $\qquad \tan 270° = \frac{-a}{0} = \infty.$ $\qquad \sec 270° = \frac{a}{0} = \infty.$

$\cos 270° = \frac{0}{a} = 0.$ $\qquad \cot 270° = \frac{0}{-a} = 0.$ $\qquad \csc 270° = \frac{a}{-a} = -1.$

Note. No absolute meaning can be attached to such a result as $\cot 0° = \infty$; it merely signifies that as an angle approaches 0°, its cotangent increases without limit.

A similar interpretation must be given to the equations $\csc 0° = \infty$, $\tan 90° = \infty$, etc.

26. The results of the last four articles may be conveniently expressed in tabular form as follows:

Angle.	*Sin.*	*Cos.*	*Tan.*	*Cot.*	*Sec.*	*Csc.*
0°	0	1	0	∞	1	∞
90°	1	0	∞	0	∞	1
180°	0	-1	0	∞	-1	∞
270°	-1	0	∞	0	∞	-1
360°	0	1	0	∞	1	∞

27. Functions of 120°, 135°, 150°, etc.

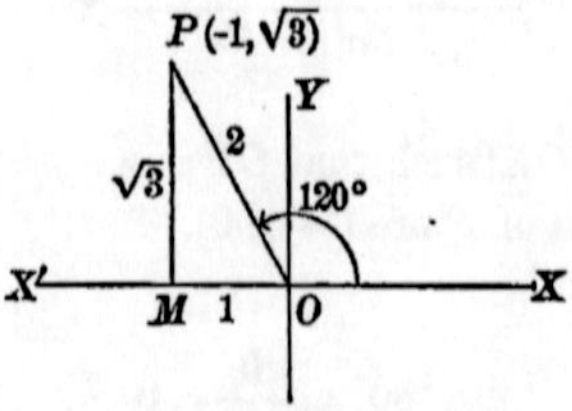

Let OPM be a right triangle having OP, OM, and PM equal to 2, 1, and $\sqrt{3}$, respectively, and $\angle POM = 60°$. (Compare § 8.)

Then $\angle XOP = 120°$, and the co-ordinates of P are $(-1, \sqrt{3})$.

Whence by definition,

$$\sin 120° = \frac{\sqrt{3}}{2}. \qquad \tan 120° = -\sqrt{3}. \qquad \sec 120° = -2.$$

$$\cos 120° = -\frac{1}{2}. \qquad \cot 120° = -\frac{1}{\sqrt{3}} = -\frac{1}{3}\sqrt{3}. \qquad \csc 120° = \frac{2}{\sqrt{3}} = \frac{2}{3}\sqrt{3}.$$

In like manner may be proved the remaining values given in the following table, which are left as exercises for the student:

Angle.	*Sin.*	*Cos.*	*Tan.*	*Cot.*	*Sec.*	*Csc.*
120°	$\frac{1}{2}\sqrt{3}$	$-\frac{1}{2}$	$-\sqrt{3}$	$-\frac{1}{3}\sqrt{3}$	-2	$\frac{2}{3}\sqrt{3}$
135°	$\frac{1}{2}\sqrt{2}$	$-\frac{1}{2}\sqrt{2}$	-1	-1	$-\sqrt{2}$	$\sqrt{2}$
150°	$\frac{1}{2}$	$-\frac{1}{2}\sqrt{3}$	$-\frac{1}{3}\sqrt{3}$	$-\sqrt{3}$	$-\frac{2}{3}\sqrt{3}$	2
210°	$-\frac{1}{2}$	$-\frac{1}{2}\sqrt{3}$	$\frac{1}{3}\sqrt{3}$	$\sqrt{3}$	$-\frac{2}{3}\sqrt{3}$	-2
225°	$-\frac{1}{2}\sqrt{2}$	$-\frac{1}{2}\sqrt{2}$	1	1	$-\sqrt{2}$	$-\sqrt{2}$
240°	$-\frac{1}{2}\sqrt{3}$	$-\frac{1}{2}$	$\sqrt{3}$	$\frac{1}{3}\sqrt{3}$	-2	$-\frac{2}{3}\sqrt{3}$
300°	$-\frac{1}{2}\sqrt{3}$	$\frac{1}{2}$	$-\sqrt{3}$	$-\frac{1}{3}\sqrt{3}$	2	$-\frac{2}{3}\sqrt{3}$
315°	$-\frac{1}{2}\sqrt{2}$	$\frac{1}{2}\sqrt{2}$	-1	-1	$\sqrt{2}$	$-\sqrt{2}$
330°	$-\frac{1}{2}$	$\frac{1}{2}\sqrt{3}$	$-\frac{1}{3}\sqrt{3}$	$-\sqrt{3}$	$\frac{2}{3}\sqrt{3}$	-2

28. *Given the value of one function of an angle, to find the values of the remaining functions.* (Compare § 6.)

1. Given $\sin A = -\frac{3}{5}$; find the values of the remaining functions of A.

The example may be solved by a method similar to that of § 6; since the sine is the ratio of the ordinate to the distance, we may regard the point of reference as having its ordinate equal to -3, and its distance equal to 5.

There are *two* points, P and P', which are 3 units below the axis of X, and distant 5 units from O.

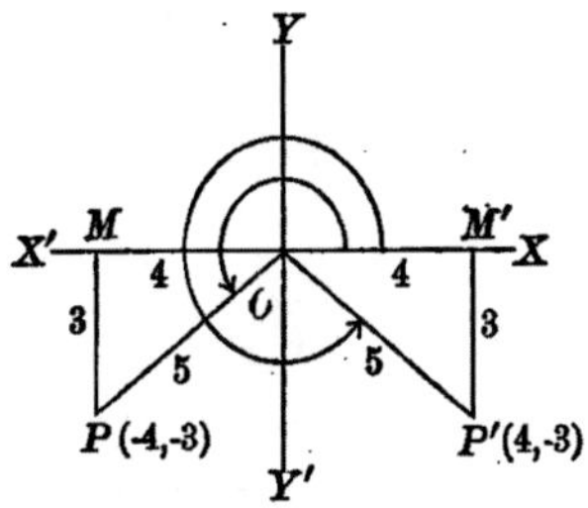

There are then *two angles*, XOP and XOP', in the third and fourth quadrants, respectively, either of which may be the angle A.

Now, $OM = OM' = \sqrt{\overline{OP}^2 - \overline{PM}^2} = \sqrt{25-9} = 4.$

Then the co-ordinates of P are $(-4, -3)$; and of P', $(4, -3)$.

Whence by definition:

Angle.	*Cos.*	*Tan.*	*Cot.*	*Sec.*	*Csc.*
XOP	$-\frac{4}{5}$	$\frac{3}{4}$	$\frac{4}{3}$	$-\frac{5}{4}$	$-\frac{5}{3}$
XOP'	$\frac{4}{5}$	$-\frac{3}{4}$	$-\frac{4}{3}$	$\frac{5}{4}$	$-\frac{5}{3}$

Thus the two solutions to the problem are:

$$\cos A = \mp\frac{4}{5},\ \tan A = \pm\frac{3}{4},\ \cot A = \pm\frac{4}{3},\ \sec A = \mp\frac{5}{4},\ \csc A = -\frac{5}{3};$$

where the upper signs refer to XOP, and the lower signs to XOP'

2. Given $\cot A = 3$; find the values of the remaining functions of A.

The equation may be written either $\cot A = \frac{3}{1}$, or $\cot A = \frac{-3}{-1}$.

We may then regard the point of reference as having its abscissa equal to 3 and its ordinate equal to 1, or as having its abscissa equal to -3 and its ordinate equal to -1.

There are *two angles*, XOP and XOP', in the first and third quadrants, respectively, either of which satisfies the given condition.

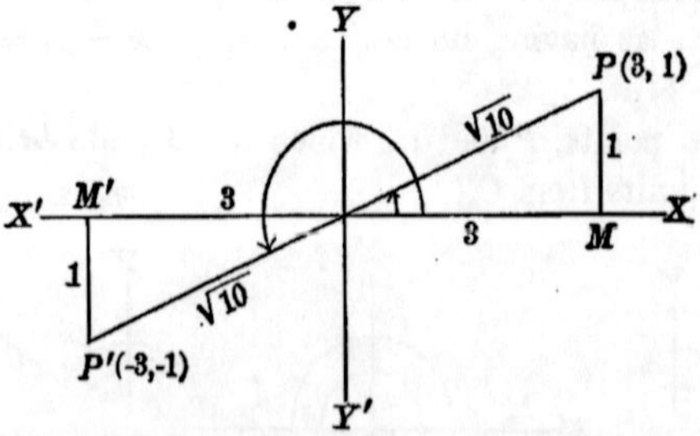

Then $OP = OP' = \sqrt{\overline{OM}^2 + \overline{PM}^2} = \sqrt{9+1} = \sqrt{10}.$

Whence by definition:

Angle.	*Sin.*	*Cos.*	*Tan.*	*Sec.*	*Csc.*
XOP	$\frac{1}{\sqrt{10}}$	$\frac{3}{\sqrt{10}}$	$\frac{1}{3}$	$\frac{\sqrt{10}}{3}$	$\sqrt{10}$
XOP'	$-\frac{1}{\sqrt{10}}$	$-\frac{3}{\sqrt{10}}$	$\frac{1}{3}$	$-\frac{\sqrt{10}}{3}$	$-\sqrt{10}$

Thus the two solutions are:

$$\sin A = \pm\frac{1}{\sqrt{10}},\ \cos A = \pm\frac{3}{\sqrt{10}},\ \tan A = \frac{1}{3},\ \sec A = \pm\frac{\sqrt{10}}{3},\ \csc A = \pm\sqrt{10}.$$

Note. It must be clearly borne in mind, in examples like the above, that the "distance" is *always positive.*

EXAMPLES.

In each of the following, find the values of the remaining functions:

3. $\sec A = \frac{5}{4}$.
4. $\cot A = -\frac{12}{5}$.
5. $\sin A = \frac{15}{17}$.
6. $\cos A = -\frac{21}{29}$.
7. $\csc A = -\frac{25}{7}$.
8. $\tan A = \frac{9}{40}$.
9. $\sec A = -\frac{7}{2}$.
10. $\sin A = -\frac{1}{5}$.
11. $\tan A = -7$.
12. $\csc A = 3$.
13. $\cos A = \frac{a}{b}$.
14. $\cot A = x$.

29. Functions of $(-A)$ in terms of those of A.

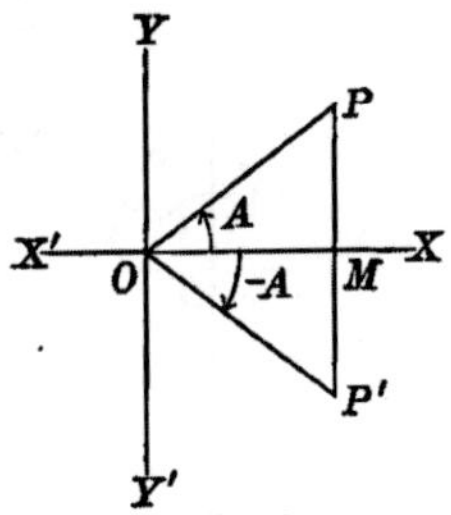

FIG. 1.

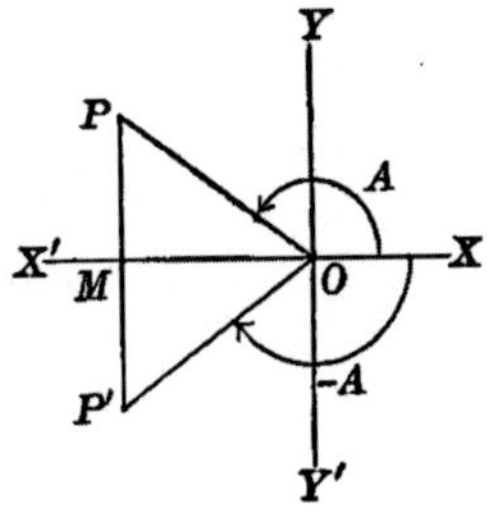

FIG. 2.

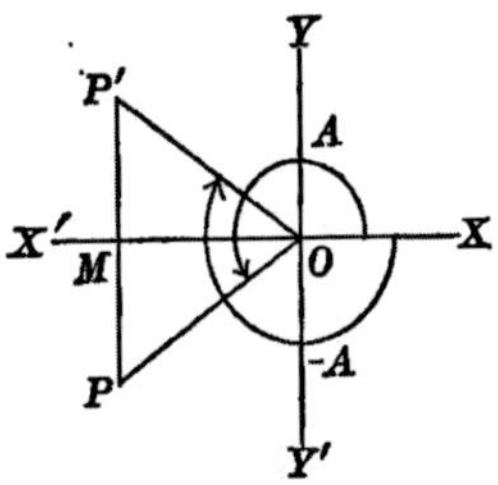

FIG. 3.

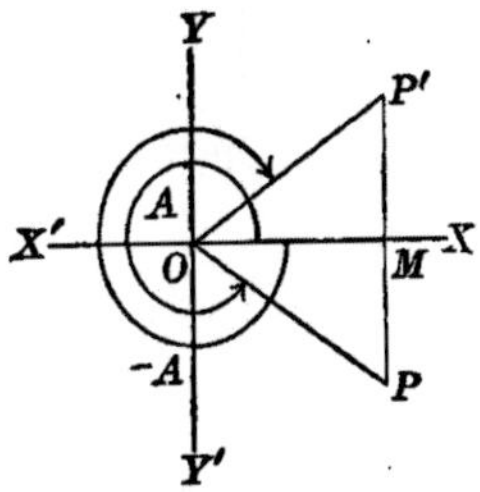

FIG. 4.

There may be four cases: A in the first quadrant (Fig. 1), A in the second quadrant (Fig. 2), A in the third quadrant (Fig. 3), or A in the fourth quadrant (Fig. 4).

In each figure, let the positive angle XOP represent the angle A, and the negative angle XOP' the angle $-A$.

Draw PM perpendicular to XX', and produce it to meet OP' at P'.

In the right triangles OPM and $OP'M$, the side OM is common, and $\angle POM = \angle P'OM$.

Hence, the triangles are equal, and $PM = P'M$ and $OP = OP'$.

Then in each figure,

$$\text{abscissa } P' = \text{abscissa } P,$$
$$\text{ordinate } P' = -\text{ordinate } P,$$

and

$$\text{distance } P' = \text{distance } P.$$

Then,

$$\frac{\text{ord. } P'}{\text{dist. } P'} = -\frac{\text{ord. } P}{\text{dist. } P}. \quad \frac{\text{ord. } P'}{\text{abs. } P'} = -\frac{\text{ord. } P}{\text{abs. } P}. \quad \frac{\text{dist. } P'}{\text{abs. } P'} = \frac{\text{dist. } P}{\text{abs. } P}.$$

$$\frac{\text{abs. } P'}{\text{dist. } P'} = \frac{\text{abs. } P}{\text{dist. } P}. \quad \frac{\text{abs. } P'}{\text{ord. } P'} = -\frac{\text{abs. } P}{\text{ord. } P}. \quad \frac{\text{dist. } P'}{\text{ord. } P'} = -\frac{\text{dist. } P}{\text{ord. } P}.$$

Whence,

$$\left.\begin{array}{lll} \sin(-A) = -\sin A. & \tan(-A) = -\tan A. & \sec(-A) = \sec A. \\ \cos(-A) = \cos A. & \cot(-A) = -\cot A. & \csc(-A) = -\csc A. \end{array}\right\} \quad (1)$$

30. Functions of $(90° + A)$ in terms of those of A.

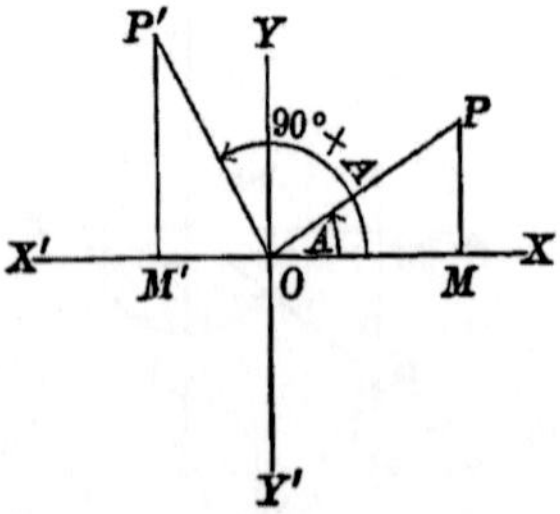

FIG. 1.

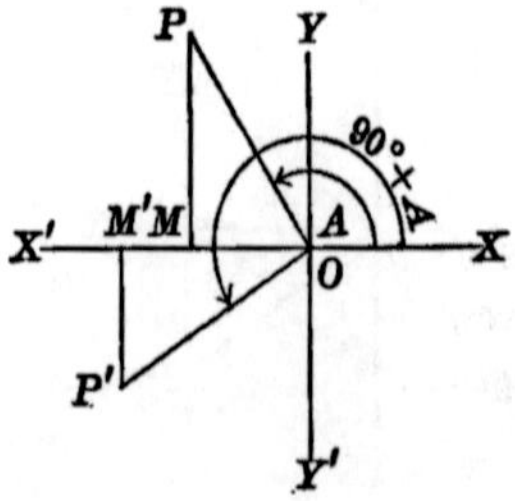

FIG. 2.

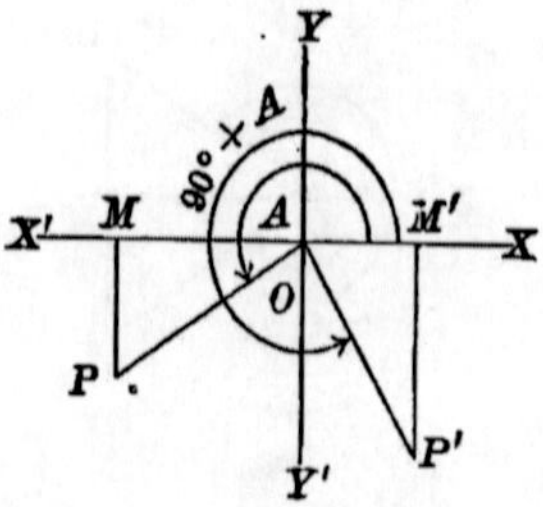

FIG. 3.

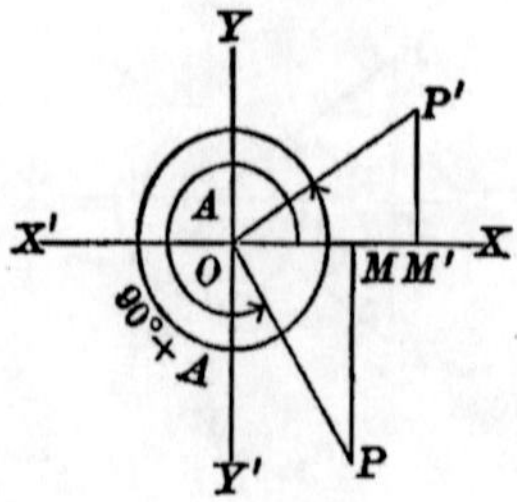

FIG. 4.

There may be four cases: A in the first quadrant (Fig. 1), A in the second quadrant (Fig. 2), A in the third quadrant (Fig. 3), or A in the fourth quadrant (Fig. 4).

In each figure, let the positive angle XOP represent the angle A, and the positive angle XOP' the angle $90° + A$.

Take $OP' = OP$, and draw PM and $P'M'$ perpendicular to XX'.

Since OP is perpendicular to OP', and OM to $P'M'$, $\angle POM = \angle OP'M'$.

Then the right triangles OPM and $OP'M'$ have the hypotenuse and an acute angle of one equal to the hypotenuse and an acute angle of the other.

Hence, the triangles are equal, and $PM = OM'$ and $OM = P'M'$.

Then in each figure,

$$\text{ordinate } P' = \text{abscissa } P,$$
$$\text{abscissa } P' = -\text{ordinate } P,$$

and

$$\text{distance } P' = \text{distance } P.$$

Then, $\dfrac{\text{ord. } P'}{\text{dist. } P'} = \dfrac{\text{abs. } P}{\text{dist. } P}$. $\qquad \dfrac{\text{abs. } P'}{\text{ord. } P'} = -\dfrac{\text{ord. } P}{\text{abs. } P}$.

$$\frac{\text{abs. } P'}{\text{dist. } P'} = -\frac{\text{ord. } P}{\text{dist. } P}. \qquad \frac{\text{dist. } P'}{\text{abs. } P'} = -\frac{\text{dist. } P}{\text{ord. } P}.$$

$$\frac{\text{ord. } P'}{\text{abs. } P'} = -\frac{\text{abs. } P}{\text{ord. } P}. \qquad \frac{\text{dist. } P'}{\text{ord. } P'} = \frac{\text{dist. } P}{\text{abs. } P}.$$

Or,
$$\left.\begin{array}{ll} \sin(90^\circ + A) = \cos A. & \cot(90^\circ + A) = -\tan A. \\ \cos(90^\circ + A) = -\sin A. & \sec(90^\circ + A) = -\csc A. \\ \tan(90^\circ + A) = -\cot A. & \csc(90^\circ + A) = \sec A. \end{array}\right\} \quad (2)$$

31. The results of § 30 may be stated as follows:

The sine, cosine, tangent, cotangent, secant, and cosecant of any angle are equal, respectively, to the cosine, minus the sine, minus the cotangent, minus the tangent, minus the cosecant, and the secant, of an angle 90° *less.*

32. Functions of $(90^\circ - A)$ **in terms of those of** A.

By § 31,
$$\begin{aligned} \sin(90^\circ - A) &= \cos(-A) = \cos A \text{ (§ 29)}. \\ \cos(90^\circ - A) &= -\sin(-A) = \sin A. \\ \tan(90^\circ - A) &= -\cot(-A) = \cot A. \\ \cot(90^\circ - A) &= -\tan(-A) = \tan A. \\ \sec(90^\circ - A) &= -\csc(-A) = \csc A. \\ \csc(90^\circ - A) &= \sec(-A) = \sec A. \end{aligned}$$

These formulæ were proved for acute angles in § 5.

33. Functions of $(180^\circ - A)$ **in terms of those of** A.

By § 31,
$$\begin{aligned} \sin(180^\circ - A) &= \cos(90^\circ - A) = \sin A \text{ (§ 32)}. \\ \cos(180^\circ - A) &= -\sin(90^\circ - A) = -\cos A. \\ \tan(180^\circ - A) &= -\cot(90^\circ - A) = -\tan A. \\ \cot(180^\circ - A) &= -\tan(90^\circ - A) = -\cot A. \\ \sec(180^\circ - A) &= -\csc(90^\circ - A) = -\sec A. \\ \csc(180^\circ - A) &= \sec(90^\circ - A) = \csc A. \end{aligned}$$

34. By successive applications of the theorem of § 31, any function of a multiple of 90°, plus or minus A, may be expressed as a function of A.

1. Express $\sin(270^\circ + A)$ as a function of A.

By § 31, $\sin(270^\circ + A) = \cos(180^\circ + A) = -\sin(90^\circ + A) = -\cos A$.

If the multiple of 90° is greater than 270°, we may subtract 360°, or any multiple of 360°, from the angle, in accordance with § 21.

2. Express $\sec(990^\circ - A)$ as a function of A.

Subtracting twice 360°, or 720°, from the angle, we have

$$\sec(990^\circ - A) = \sec(270^\circ - A).$$

And by § 31, $\sec(270^\circ - A) = -\csc(180^\circ - A) = -\csc A$ (§ 33).

If the multiple of 90° is negative, we may add 360°, or any multiple of 360°, to the angle.

3. Express $\tan(-180° + A)$ as a function of A.

Adding 360° to the angle, we have

$$\tan(-180° + A) = \tan(180° + A).$$

And by § 31, $\tan(180° + A) = -\cot(90° + A) = \tan A.$

EXAMPLES.

Express each of the following as a function of A:

4. $\sin(180° + A)$.
5. $\cos(270° - A)$.
6. $\cot(450° + A)$.
7. $\csc(360° - A)$.
8. $\tan(540° - A)$.
9. $\sec(630° + A)$.
10. $\tan(-270° - A)$.
11. $\csc(-90° - A)$.
12. $\cot(-180° + A)$.
13. $\sin(-630° + A)$.
14. $\tan(-450° - A)$.
15. $\cos(-900° - A)$.
16. $\sin(810° - A)$.
17. $\csc(1080° + A)$.
18. $\sec(1260° + A)$.

35. By means of the theorem of § 31, any function of any angle, positive or negative, may be expressed as a function of a certain acute angle.

1. Express $\sin 317°$ as a function of an acute angle.

By § 31, $\sin 317° = \cos 227° = -\sin 137° = -\cos 47°$.

Since the complement of 47° is 43°, another form of the result is $-\sin 43°$ (§ 5).

Note. As in the examples of § 34, 360°, or any multiple of 360°, may be added to, or subtracted from, the angle.

EXAMPLES.

Express each of the following as a function of an acute angle:

2. $\cos 322°$.
3. $\tan 208°$.
4. $\sec 559°$.
5. $\csc 803° 45'$.
6. $\cot(-378°)$.
7. $\sin(-139° 5')$.

It is evident from the above that any function of any angle can be expressed as a function of a certain acute angle *less than* 45°.

Express each of the following as a function of an acute angle less than 45°:

8. $\cot 155°$.
9. $\sin 1138° 36'$.
10. $\sec 457°$.
11. $\cos 496° 20'$.
12. $\tan(-681°)$.
13. $\csc(-257°)$.

14. Find the value of $\csc(-210°)$.

Adding 360° to the angle, we have

$$\csc(-210°) = \csc 150°.$$

And by § 31, $\csc 150° = \sec 60° = 2$ (§ 8).

Find the values of the following:

15. $\cot 405°$.
16. $\sin 480°$.
17. $\csc 600°$.
18. $\tan 690°$.
19. $\cos(-420°)$.
20. $\sec(-225°)$.

III. GENERAL FORMULÆ.

36. It follows immediately from the definitions of § 17 that, if x is any angle,

$$\left.\begin{array}{lll} \sin x = \dfrac{1}{\csc x}. & \tan x = \dfrac{1}{\cot x}. & \sec x = \dfrac{1}{\cos x}. \\ \cos x = \dfrac{1}{\sec x}. & \cot x = \dfrac{1}{\tan x}. & \csc x = \dfrac{1}{\sin x}. \end{array}\right\} \quad (3)$$

37. *To prove the formula*

$$\tan x = \frac{\sin x}{\cos x}. \quad (4)$$

I. When the angle x is acute (or in the first quadrant).

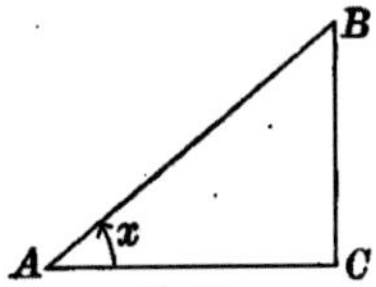

Fig. 1.

In the right triangle ABC, let BAC be the angle x.

By § 2,
$$\tan x = \frac{BC}{AC} = \frac{\dfrac{BC}{AB}}{\dfrac{AC}{AB}} = \frac{\sin x}{\cos x}.$$

II. When x is in the second, third, or fourth quadrant.

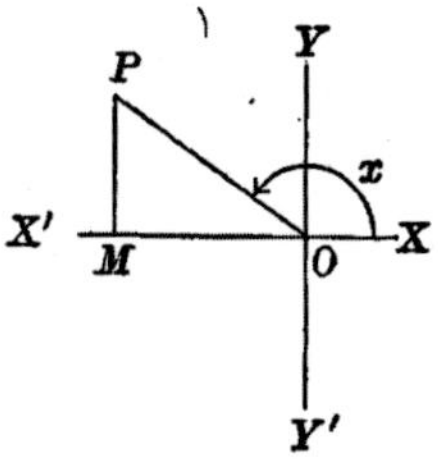

Fig. 2.

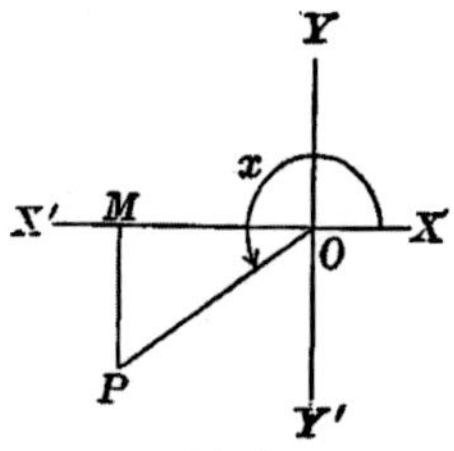

Fig. 3.

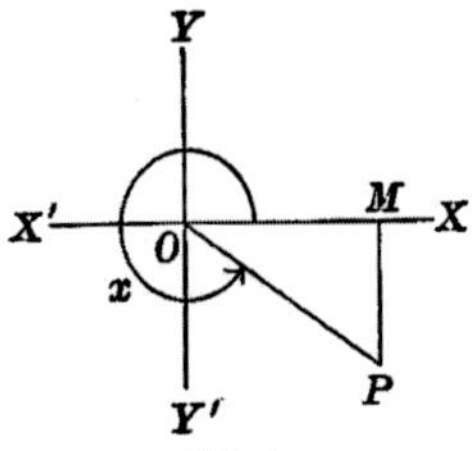

Fig. 4.

In each figure, let the positive angle XOP represent the angle x, and draw PM perpendicular to XX'.

Then in each figure, by the definitions of § 17,

$$\tan x = \frac{\text{ord. } P}{\text{abs. } P} = \frac{\dfrac{\text{ord. } P}{\text{dist. } P}}{\dfrac{\text{abs. } P}{\text{dist. } P}} = \frac{\sin x}{\cos x}.$$

38. *To prove the formula*

$$\cot x = \frac{\cos x}{\sin x}. \qquad (5)$$

By (3), § 36, $\cot x = \frac{1}{\tan x} = \frac{1}{\frac{\sin x}{\cos x}}$ (§ 37) $= \frac{\cos x}{\sin x}.$

39. *To prove the formula*

$$\sin^2 x + \cos^2 x = 1. \qquad (6)$$

Note. $\text{Sin}^2 x$ signifies $(\sin x)^2$; that is, the square of the sine of x.

I. When the angle x is acute (or in the first quadrant).

In Fig. 1, § 37, we have by Geometry,

$$\overline{BC}^2 + \overline{AC}^2 = \overline{AB}^2.$$

Dividing by $\overline{AB}^2$, $\left(\frac{BC}{AB}\right)^2 + \left(\frac{AC}{AB}\right)^2 = 1.$

Then by definition, $(\sin x)^2 + (\cos x)^2 = 1.$

That is, $\sin^2 x + \cos^2 x = 1.$

II. When x is in the second, third, or fourth quadrant.

In each of the figures 2, 3, and 4 of § 37, we have

$$\overline{PM}^2 + \overline{OM}^2 = \overline{OP}^2.$$

Dividing by $\overline{OP}^2$, $\frac{\overline{PM}^2}{\overline{OP}^2} + \frac{\overline{OM}^2}{\overline{OP}^2} = 1.$

But in either figure, $\frac{\overline{PM}^2}{\overline{OP}^2} = \sin^2 x$, and $\frac{\overline{OM}^2}{\overline{OP}^2} = \cos^2 x.$

Whence, $\sin^2 x + \cos^2 x = 1.$

Formula (6) may be written in the forms

$$\sin^2 x = 1 - \cos^2 x, \text{ and } \cos^2 x = 1 - \sin^2 x.$$

40. *To prove the formulæ*

$$\sec^2 x = 1 + \tan^2 x, \qquad (7)$$

and

$$\csc^2 x = 1 + \cot^2 x. \qquad (8)$$

By (6), $1 = \cos^2 x + \sin^2 x. \qquad$ (A)

Dividing by $\cos^2 x$, $\frac{1}{\cos^2 x} = 1 + \frac{\sin^2 x}{\cos^2 x}.$

Whence by (3) and (4), $\sec^2 x = 1 + \tan^2 x.$

Again, dividing (A) by $\sin^2 x$, we have

$$\frac{1}{\sin^2 x} = 1 + \frac{\cos^2 x}{\sin^2 x}.$$

Whence by (3) and (5), $\csc^2 x = 1 + \cot^2 x$.

41. *To find the values of* $\sin(x+y)$ *and* $\cos(x+y)$ *in terms of the sines and cosines of* x *and* y.

I. When x and y are acute, and $x+y$ acute.

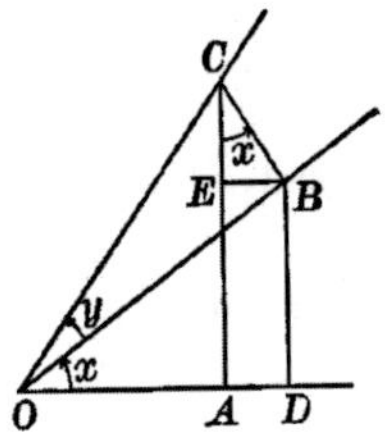

Let AOB and BOC denote the angles x and y, respectively.

Then, $\angle AOC = x + y$.

From any point C in OC draw CA and CB perpendicular to OA and OB; and draw BD and BE perpendicular to OA and AC.

Since EC and BC are perpendicular to OA and OB, the angles BCE and AOB are equal; that is, $\angle BCE = x$.

Now, $$\sin(x+y) = \frac{AC}{OC} = \frac{BD + CE}{OC} = \frac{BD}{OC} + \frac{CE}{OC}.$$

But, $$\frac{BD}{OC} = \frac{BD}{OB} \times \frac{OB}{OC} = \sin x \cos y,$$

and $$\frac{CE}{OC} = \frac{CE}{BC} \times \frac{BC}{OC} = \cos x \sin y.$$

Whence, $$\sin(x+y) = \sin x \cos y + \cos x \sin y. \quad (9)$$

Again, $$\cos(x+y) = \frac{OA}{OC} = \frac{OD - BE}{OC} = \frac{OD}{OC} - \frac{BE}{OC}.$$

But, $$\frac{OD}{OC} = \frac{OD}{OB} \times \frac{OB}{OC} = \cos x \cos y,$$

and $$\frac{BE}{OC} = \frac{BE}{BC} \times \frac{BC}{OC} = \sin x \sin y.$$

Whence, $$\cos(x+y) = \cos x \cos y - \sin x \sin y. \quad (10)$$

II. When x and y are acute, and $x+y$ obtuse.

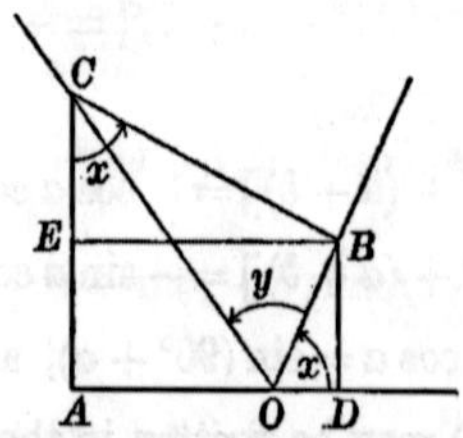

Let DOB and BOC denote the angles x and y, respectively.

Then, $\angle DOC = x + y$.

From any point C in OC draw CB perpendicular to OB, and CA perpendicular to DO produced; and draw BD and BE perpendicular to OD and AC.

Since EC and BC are perpendicular to OD and OB, the angles BCE and DOB are equal; that is, $\angle BCE = x$.

Then by § 17, $\sin DOC = \dfrac{AC}{OC} = \dfrac{BD + CE}{OC} = \dfrac{BD}{OC} + \dfrac{CE}{OC}$.

But,
$$\frac{BD}{OC} = \frac{BD}{OB} \times \frac{OB}{OC} = \sin x \cos y,$$

and
$$\frac{CE}{OC} = \frac{CE}{BC} \times \frac{BC}{OC} = \cos x \sin y.$$

Whence, $\sin(x+y) = \sin x \cos y + \cos x \sin y.$

Again,
$$\cos DOC = \frac{-OA}{OC} = \frac{OD - BE}{OC} = \frac{OD}{OC} - \frac{BE}{OC}.$$

But,
$$\frac{OD}{OC} = \frac{OD}{OB} \times \frac{OB}{OC} = \cos x \cos y,$$

and
$$\frac{BE}{OC} = \frac{BE}{BC} \times \frac{BC}{OC} = \sin x \sin y.$$

Whence, $\cos(x+y) = \cos x \cos y - \sin x \sin y.$

42. Formulæ (**9**) and (**10**) are very important, and it is necessary to prove them *for all values of x and y.*

They have already been proved when x and y are any two acute angles; or, what is the same thing, when they are any two angles in the *first quadrant.*

Now let a and b be any assigned values of x and y, for which (**9**) and (**10**) are true.

By (2), § 30, $\sin[90° + (a+b)] = \cos(a+b)$,

and $\cos[90° + (a+b)] = -\sin(a+b)$.

Whence, by (9) and (10),

$$\sin[90° + (a+b)] = \cos a \cos b - \sin a \sin b, \qquad \text{(A)}$$

and

$$\cos[90° + (a+b)] = -\sin a \cos b - \cos a \sin b. \qquad \text{(B)}$$

But by (2), § 30, $\cos a = \sin(90° + a)$, and $-\sin a = \cos(90° + a)$.

Then, (A) and (B) may be written in the forms

$$\sin[(90° + a) + b] = \sin(90° + a)\cos b + \cos(90° + a)\sin b,$$

and

$$\cos[(90° + a) + b] = \cos(90° + a)\cos b - \sin(90° + a)\sin b;$$

which are in accordance with (9) and (10).

Therefore, if (9) and (10) hold for any assigned values of x and y, they also hold when one of the angles is increased by 90°.

But they have been proved to hold when x and y are both in the first quadrant; hence, they hold when x is in the second quadrant and y in the first.

And since they hold when x is in the second quadrant and y in the first, they hold when x and y are both in the second quadrant; and so on.

Hence, (9) and (10) hold for any values of x and y whatever, positive or negative.

43. By (9), $\sin[x + (-y)] = \sin x \cos(-y) + \cos x \sin(-y)$

$= \sin x \cos y + \cos x(-\sin y)$, by (1), § 29.

Whence,

$$\sin(x - y) = \sin x \cos y - \cos x \sin y. \qquad \textbf{(11)}$$

By (10), $\cos[x + (-y)] = \cos x \cos(-y) - \sin x \sin(-y)$

$= \cos x \cos y - \sin x(-\sin y)$.

Whence,

$$\cos(x - y) = \cos x \cos y + \sin x \sin y. \qquad \textbf{(12)}$$

44. By (4),

$$\tan(x+y) = \frac{\sin(x+y)}{\cos(x+y)}$$

$$= \frac{\sin x \cos y + \cos x \sin y}{\cos x \cos y - \sin x \sin y}, \text{ by (9) and (10).}$$

Dividing each term of the fraction by $\cos x \cos y$,

$$\tan(x+y) = \frac{\dfrac{\sin x \cos y}{\cos x \cos y} + \dfrac{\cos x \sin y}{\cos x \cos y}}{\dfrac{\cos x \cos y}{\cos x \cos y} - \dfrac{\sin x \sin y}{\cos x \cos y}}$$

$$= \frac{\tan x + \tan y}{1 - \tan x \tan y}. \qquad \textbf{(13)}$$

In like manner, we may prove

$$\tan(x-y)=\frac{\tan x-\tan y}{1+\tan x\tan y}. \tag{14}$$

Again, by (5), $\cot(x+y)=\dfrac{\cos(x+y)}{\sin(x+y)}$

$$=\frac{\cos x\cos y-\sin x\sin y}{\sin x\cos y+\cos x\sin y}.$$

Dividing each term of the fraction by $\sin x\sin y$,

$$\cot(x+y)=\frac{\dfrac{\cos x\cos y}{\sin x\sin y}-\dfrac{\sin x\sin y}{\sin x\sin y}}{\dfrac{\sin x\cos y}{\sin x\sin y}+\dfrac{\cos x\sin y}{\sin x\sin y}}$$

$$=\frac{\cot x\cot y-1}{\cot y+\cot x}. \tag{15}$$

In like manner, we may prove

$$\cot(x-y)=\frac{\cot x\cot y+1}{\cot y-\cot x}. \tag{16}$$

45. From (9), (10), (11), and (12), we have

$$\sin(a+b)=\sin a\cos b+\cos a\sin b. \tag{A}$$

$$\sin(a-b)=\sin a\cos b-\cos a\sin b. \tag{B}$$

$$\cos(a+b)=\cos a\cos b-\sin a\sin b. \tag{C}$$

$$\cos(a-b)=\cos a\cos b+\sin a\sin b. \tag{D}$$

Adding and subtracting (A) and (B), and then (C) and (D),

$$\sin(a+b)+\sin(a-b)=2\sin a\cos b.$$

$$\sin(a+b)-\sin(a-b)=2\cos a\sin b.$$

$$\cos(a+b)+\cos(a-b)=2\cos a\cos b.$$

$$\cos(a+b)-\cos(a-b)=-2\sin a\sin b.$$

Let $a+b=x$, and $a-b=y$.
Then, $a=\frac{1}{2}(x+y)$, and $b=\frac{1}{2}(x-y)$.
Substituting these values, we have

$$\sin x+\sin y=2\sin\tfrac{1}{2}(x+y)\cos\tfrac{1}{2}(x-y). \tag{17}$$

$$\sin x-\sin y=2\cos\tfrac{1}{2}(x+y)\sin\tfrac{1}{2}(x-y). \tag{18}$$

$$\cos x+\cos y=2\cos\tfrac{1}{2}(x+y)\cos\tfrac{1}{2}(x-y). \tag{19}$$

$$\cos x-\cos y=-2\sin\tfrac{1}{2}(x+y)\sin\tfrac{1}{2}(x-y). \tag{20}$$

46. By (**17**) and (**18**), we have

$$\frac{\sin x+\sin y}{\sin x-\sin y}=\frac{2\sin\frac{1}{2}(x+y)\cos\frac{1}{2}(x-y)}{2\cos\frac{1}{2}(x+y)\sin\frac{1}{2}(x-y)}$$
$$=\tan\tfrac{1}{2}(x+y)\cot\tfrac{1}{2}(x-y)$$
$$=\frac{\tan\frac{1}{2}(x+y)}{\tan\frac{1}{2}(x-y)},\text{ by (3).} \qquad (21)$$

47. By (**9**) and (**11**), we have

$$\sin(x+y)\sin(x-y)=(\sin x\cos y+\cos x\sin y)(\sin x\cos y-\cos x\sin y)$$
$$=\sin^2x\cos^2y-\cos^2x\sin^2y$$
$$=\sin^2x(1-\sin^2y)-(1-\sin^2x)\sin^2y \quad (\S\ 39)$$
$$=\sin^2x-\sin^2x\sin^2y-\sin^2y+\sin^2x\sin^2y$$
$$=\sin^2x-\sin^2y. \qquad (22)$$

The result may also be written

$$\sin(x+y)\sin(x-y)=1-\cos^2x-(1-\cos^2y) \quad (\S\ 39)$$
$$=\cos^2y-\cos^2x. \qquad (23)$$

In like manner, we may prove

$$\cos(x+y)\cos(x-y)=\cos^2x-\sin^2y=\cos^2y-\sin^2x. \qquad (24)$$

48. Functions of $2x$.

Putting $y=x$ in (**9**), we have

$$\sin 2x=\sin x\cos x+\cos x\sin x$$
$$=2\sin x\cos x. \qquad (25)$$

Putting $y=x$ in (**10**), we obtain

$$\cos 2x=\cos^2x-\sin^2x. \qquad (26)$$

We also have by § 39,

$$\cos 2x=(1-\sin^2x)-\sin^2x=1-2\sin^2x, \qquad (27)$$

and

$$\cos 2x=\cos^2x-(1-\cos^2x)=2\cos^2x-1. \qquad (28)$$

Putting $y=x$ in (**13**) and (**15**), we have

$$\tan 2x=\frac{2\tan x}{1-\tan^2x}, \qquad (29)$$

$$\cot 2x=\frac{\cot^2x-1}{2\cot x}. \qquad (30)$$

49. Functions of $\frac{1}{2}x$.

From (**27**) and (**28**) we have, by transposition,

$$2\sin^2x=1-\cos 2x,\text{ and }2\cos^2x=1+\cos 2x.$$

Putting $\frac{1}{2}x$ in place of x, and therefore x in place of $2x$, we have

$$2\sin^2\tfrac{1}{2}x = 1 - \cos x, \tag{31}$$

$$2\cos^2\tfrac{1}{2}x = 1 + \cos x. \tag{32}$$

Again, putting $\frac{1}{2}x$ in place of x in (**25**),

$$2\sin\tfrac{1}{2}x\cos\tfrac{1}{2}x = \sin x. \tag{A}$$

Dividing (**31**) by (A), we have, by (**4**),

$$\tan\tfrac{1}{2}x = \frac{1-\cos x}{\sin x}. \tag{33}$$

Dividing (**32**) by (A), $\quad \cot\frac{1}{2}x = \dfrac{1+\cos x}{\sin x}.$ (**34**)

50. Functions of $3x$.

We have, $\sin 3x = \sin(2x+x) = \sin 2x\cos x + \cos 2x\sin x$, by (**9**)

$$\begin{aligned} &= (2\sin x\cos x)\cos x + (1-2\sin^2 x)\sin x \ (\S\ 48) \\ &= 2\sin x(1-\sin^2 x) + \sin x - 2\sin^3 x \ (\S\ 39) \\ &= 3\sin x - 4\sin^3 x. \end{aligned} \tag{35}$$

Also, $\quad \cos 3x = \cos(2x+x) = \cos 2x\cos x - \sin 2x\sin x$, by (**10**)

$$\begin{aligned} &= (2\cos^2 x - 1)\cos x - (2\sin x\cos x)\sin x \ (\S\ 48) \\ &= 2\cos^3 x - \cos x - 2\cos x(1-\cos^2 x) \ (\S\ 39) \\ &= 4\cos^3 x - 3\cos x. \end{aligned} \tag{36}$$

Again, $\quad \tan 3x = \tan(2x+x) = \dfrac{\tan 2x + \tan x}{1 - \tan 2x\tan x}$, by (**13**)

$$= \frac{\dfrac{2\tan x}{1-\tan^2 x} + \tan x}{1 - \left(\dfrac{2\tan x}{1-\tan^2 x}\right)\tan x}, \text{ by } (\mathbf{29})$$

$$= \frac{2\tan x + (1-\tan^2 x)\tan x}{1 - \tan^2 x - 2\tan^2 x} = \frac{3\tan x - \tan^3 x}{1 - 3\tan^2 x}. \tag{37}$$

EXERCISES.

51. 1. Prove the relation $\sec^2 x\csc^2 x = \sec^2 x + \csc^2 x$.

By (**3**), $\sec^2 x\csc^2 x = \dfrac{1}{\cos^2 x\sin^2 x} = \dfrac{\sin^2 x + \cos^2 x}{\cos^2 x\sin^2 x}$, by (**6**)

$$= \frac{\sin^2 x}{\cos^2 x\sin^2 x} + \frac{\cos^2 x}{\cos^2 x\sin^2 x}$$

$$= \frac{1}{\cos^2 x} + \frac{1}{\sin^2 x} = \sec^2 x + \csc^2 x.$$

2. Prove the relation $\dfrac{\sin 3x - \sin x}{\cos 3x + \cos x} = \tan x.$

By (**18**) and (**19**), $\dfrac{\sin 3x - \sin x}{\cos 3x + \cos x} = \dfrac{2\cos\frac{1}{2}(3x+x)\sin\frac{1}{2}(3x-x)}{2\cos\frac{1}{2}(3x+x)\cos\frac{1}{2}(3x-x)} = \tan x.$

3. Prove the relation $\dfrac{\tan(x+y) - \tan x}{1 + \tan(x+y)\tan x} = \tan y.$

By (**14**), $\dfrac{\tan(x+y) - \tan x}{1 + \tan(x+y)\tan x} = \tan[(x+y) - x] = \tan y.$

Prove the following relations:

4. $\dfrac{\sin(x+y)}{\sin(x-y)} = \dfrac{\tan x + \tan y}{\tan x - \tan y}.$

5. $\dfrac{\cos(x+y)}{\cos(x-y)} = \dfrac{1 - \tan x \tan y}{1 + \tan x \tan y}.$

6. $\dfrac{\cos x + \cos y}{\cos x - \cos y} = -\cot\frac{1}{2}(x+y)\cot\frac{1}{2}(x-y).$

7. $\sin(x+y+z) = \sin x \cos y \cos z + \cos x \sin y \cos z$
$\qquad + \cos x \cos y \sin z - \sin x \sin y \sin z.$

8. $\cos(x+y+z) = \cos x \cos y \cos z - \sin x \sin y \cos z$
$\qquad - \sin x \cos y \sin z - \cos x \sin y \sin z.$

9. $\tan(60°+x) - \cot(30°-x) = 0.$

10. $\dfrac{\csc^2 A}{\csc^2 A - 2} = \sec 2A.$

11. $\dfrac{\sin 2x}{\sin x} - \dfrac{\cos 2x}{\cos x} = \sec x.$

12. $\left(\dfrac{\tan x + 1}{\tan x - 1}\right)^2 = \dfrac{1 + \sin 2x}{1 - \sin 2x}.$

13. $\dfrac{\sin 5x + \sin x}{\cos 5x + \cos x} = \tan 3x.$

14. $\dfrac{\sin 3x - \sin 5x}{\cos 3x - \cos 5x} = -\cot 4x.$

15. $\sin 4x = 4\sin x \cos x - 8\sin^3 x \cos x.$

16. $\cos 4x = 1 - 8\cos^2 x + 8\cos^4 x.$

17. By putting $x = 45°$ and $y = 30°$ in (**11**) and (**12**), prove

$$\sin 15° = \tfrac{1}{4}(\sqrt{6} - \sqrt{2}),\ \cos 15° = \tfrac{1}{4}(\sqrt{6} + \sqrt{2}).$$

18. By putting $x = 30°$ in (**33**) and (**34**), prove

$$\tan 15° = 2 - \sqrt{3},\ \cot 15° = 2 + \sqrt{3}.$$

19. Using the results of Ex. 17, prove

$$\sec 15° = \sqrt{6} - \sqrt{2},\ \csc 15° = \sqrt{6} + \sqrt{2}.$$

20. By putting $x = 45°$ in (**31**) and (**32**), prove

$$\sin 22\tfrac{1}{2}° = \tfrac{1}{2}\sqrt{2 - \sqrt{2}},\ \cos 22\tfrac{1}{2}° = \tfrac{1}{2}\sqrt{2 + \sqrt{2}}.$$

21. By putting $x = 45°$ in (**33**) and (**34**), prove

$$\tan 22\tfrac{1}{2}° = \sqrt{2} - 1,\ \cot 22\tfrac{1}{2} = \sqrt{2} + 1.$$

22. By putting $x = 22\tfrac{1}{2}°$ in (**7**) and (**8**), and using the result of Ex. 21, prove

$$\sec 22\tfrac{1}{2}° = \sqrt{4 - 2\sqrt{2}},\ \csc 22\tfrac{1}{2}° = \sqrt{4 + 2\sqrt{2}}.$$

Prove the following relations:

23. $\tan(45° + x) - \tan(45° - x) = 2\tan 2x.$

24. $\cos^4 x - \sin^4 x = \cos 2x.$

25. $\dfrac{1}{\csc x - \cot x} = \cot \tfrac{1}{2} x.$

26. $\sin^2(x + y) - \sin^2(x - y) = \sin 2x \sin 2y.$

27. $\dfrac{\tan(x - y) + \tan y}{1 - \tan(x - y)\tan y} = \tan x.$

28. $\cos 5A \cos 3A + \sin 5A \sin 3A = \cos 2A.$

29. $\sin(A + B)\cos(A - B) - \cos(A + B)\sin(A - B) = \sin 2B.$

30. $\dfrac{\cos 3x}{\sin x} + \dfrac{\sin 3x}{\cos x} = 2\cot 2x.$

31. $\sin 2x = \dfrac{2\tan x}{1 + \tan^2 x}.$

32. $\cos 2x = \dfrac{1 - \tan^2 x}{1 + \tan^2 x}.$

33. $\dfrac{\sin 4x + \sin 3x}{\cos 3x - \cos 4x} = \cot \tfrac{1}{2} x.$

34. $\sin 50° + \sin 10° = \sin 70°.$

35. $\dfrac{\sin x + \sin 2x}{1 + \cos x + \cos 2x} = \tan x.$

36. $2\cos 3x \sin x = \sin 4x - \sin 2x.$

37. $\cos 5x = 5\cos x - 20\cos^3 x + 16\cos^5 x.$

38. $\dfrac{\cos x}{1 - \sin x} = \dfrac{\cot \frac{1}{2} x + 1}{\cot \frac{1}{2} x - 1}.$

39. $\tan 4x = \dfrac{4\tan x - 4\tan^3 x}{1 - 6\tan^2 x + \tan^4 x}.$

40. $(\sin x + \cos x)(2 - \sin 2x) = 2(\sin^3 x + \cos^3 x).$

41. $(\sin x - \sin y)^2 + (\cos x - \cos y)^2 = 4\sin^2 \dfrac{x - y}{2}.$

42. $\dfrac{1 + \sin x - \cos x}{1 + \sin x + \cos x} = \tan \tfrac{1}{2} x.$

43. $\dfrac{\sin 3x - \cos 3x}{\sin x + \cos x} = 2\sin 2x - 1.$

IV. MISCELLANEOUS THEOREMS.

52. Circular Measure of an Angle.

An angle is measured by finding its ratio to another angle, adopted arbitrarily as the unit of measure.

The usual unit of measure for angles is the degree, which is an angle equal to the ninetieth part of a right angle.

Another method of measuring angles, and one of great importance, is known as the *Circular Method;* in which the unit of measure is *the angle at the centre of a circle subtended by an arc whose length is equal to the radius.*

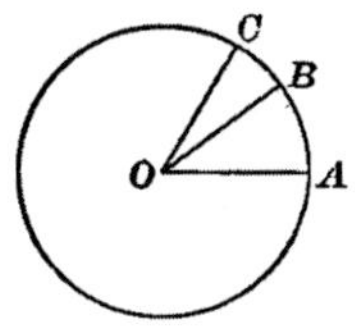

Thus, let AOB be any angle; and let AOC be the unit of circular measure; that is, the angle at the centre subtended by an arc whose length is equal to OA.

Then, $$\text{circular measure } AOB = \frac{\angle AOB}{\angle AOC}.$$

But by Geometry, $$\frac{\angle AOB}{\angle AOC} = \frac{\text{arc } AB}{\text{arc } AC} = \frac{\text{arc } AB}{OA}.$$

Whence, $$\text{circular measure } AOB = \frac{\text{arc } AB}{OA}.$$

That is, *the circular measure of an angle is the ratio of its subtending arc to the radius of the circle.*

53. By § 52, the circular measure of a right angle is the ratio of one-fourth the circumference to the radius.

But if R denotes the radius, the circumference of the circle is $2\pi R$.

Whence, $$\text{circular measure of } 90° = \frac{\frac{1}{4} \text{ of } 2\pi R}{R} = \frac{\pi}{2}.$$

It follows from the above that the circular measure of 180° is π; of 60°, $\frac{\pi}{3}$; of 45°, $\frac{\pi}{4}$; etc.

That is, *an angle expressed in degrees may be reduced to circular measure by finding its ratio to* 180°, *and multiplying the result by* π.

Thus, since 115° is $\frac{23}{36}$ of 180°, the circular measure of 115° is $\frac{23\pi}{36}$.

54. Conversely, *an angle expressed in circular measure may be reduced to degrees by multiplying by* 180° *and dividing by* π; or, more briefly, *by substituting* 180° *for* π.

Thus, $\frac{7\pi}{15} = \frac{7}{15}$ of $180° = 84°$.

55. In the circular method, such expressions may occur as "the angle $\frac{2}{3}$," "the angle 1," etc.

These refer to the unit of circular measure; thus, the angle $\frac{2}{3}$ signifies an angle whose subtending arc is two-thirds of the radius.

The angle 1, that is, the angle whose subtending arc is equal to the radius, or the unit of circular measure, reduced to degrees by the rule of § 54, gives

$$\frac{180°}{\pi} = \frac{180°}{3.14159 \cdots} = 57.2958°, \text{ approximately.}$$

Then the rule of § 54 may be modified as follows:

An angle expressed in circular measure may be reduced to degrees by multiplying by 57.2958°.

Thus, the angle $\frac{2}{3} = \frac{2}{3} \times 57.2958° = 38.1972° = 38°\,11'\,49.92''$.

EXAMPLES.

56. Express each of the following in circular measure:

1. 120°.
2. 315°.
3. 67° 30′.
4. 146° 15′.
5. 86° 24′.
6. 53° 20′.
7. 163° 7′ 30″.
8. 88° 53′ 20″.

Express each of the following in degree measure:

9. $\frac{5\pi}{6}$.
10. $\frac{11\pi}{24}$.
11. $\frac{23\pi}{64}$.
12. $\frac{3}{2}$.
13. $\frac{1}{4}$.
14. $\frac{5}{3}$.
15. $\frac{\pi - 1}{6}$.
16. $\frac{3\pi + 2}{5}$.

57. Inverse Trigonometric Functions.

The expression $\sin^{-1} x$, called the *inverse sine* of x, or the *anti-sine* of x, signifies *the angle whose sine is* x.

Thus, the statement that the sine of the angle x is equal to y may be expressed in either of the ways

$$\sin x = y, \text{ or } x = \sin^{-1} y.$$

In like manner, $\cos^{-1} x$ signifies the angle whose cosine is x; $\tan^{-1} x$, the angle whose tangent is x; etc.

Note. The student must be careful not to confuse the above notation with the *exponent* -1; the -1 power of $\sin x$ is expressed $(\sin x)^{-1}$, and not $\sin^{-1} x$.

It is evident that the sine of the angle whose sine is x is x; that is, $\sin(\sin^{-1}x) = x$.

In like manner, $\cos(\cos^{-1}x) = x$; $\tan(\tan^{-1}x) = x$; etc.

58. By aid of the principles of § 57, we may derive from any formula involving direct functions a relation between inverse functions.

1. From the formula $\tan(x+y) = \dfrac{\tan x + \tan y}{1 - \tan x \tan y}$, prove

$$\tan^{-1}a + \tan^{-1}b = \tan^{-1}\frac{a+b}{1-ab}.$$

Let $\qquad \tan x = a$, and $\tan y = b$.

Then by § 57, $\qquad x = \tan^{-1}a$, and $y = \tan^{-1}b$.

Substituting these values in the given formula,

$$\tan(\tan^{-1}a + \tan^{-1}b) = \frac{a+b}{1-ab}.$$

Whence, $\qquad \tan^{-1}a + \tan^{-1}b = \tan^{-1}\dfrac{a+b}{1-ab}.$

2. Prove the relation $\quad \cot^{-1}a - \sec^{-1}b = \cos^{-1}\dfrac{a+\sqrt{b^2-1}}{b\sqrt{a^2+1}}.$

Let $\qquad \cot^{-1}a = x$, and $\sec^{-1}b = y$.

Then, $\qquad \cot x = a$, and $\sec y = b$.

Now, $\qquad \cos(x-y) = \cos x \cos y + \sin x \sin y. \qquad$ (A)

To find the values of the sines and cosines of x and y, we may use the method of § 6.

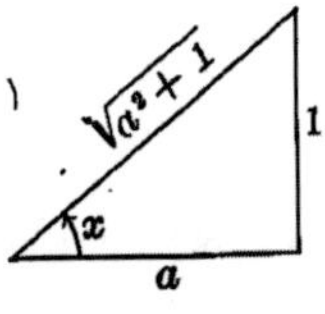

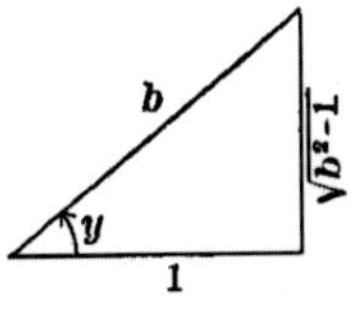

In the right triangle containing the angle x, the adjacent side is a, and the opposite side 1; then, the hypotenuse is $\sqrt{a^2+1}$.

In the right triangle containing the angle y, the hypotenuse is b, and the adjacent side 1; then, the opposite side is $\sqrt{b^2-1}$.

Substituting the values of $\cos x$, $\cos y$, $\sin x$, and $\sin y$ in (A), we have

$$\cos(x-y) = \frac{a}{\sqrt{a^2+1}} \cdot \frac{1}{b} + \frac{1}{\sqrt{a^2+1}} \cdot \frac{\sqrt{b^2-1}}{b} = \frac{a+\sqrt{b^2-1}}{b\sqrt{a^2+1}}.$$

Whence, $\quad x - y$ or $\cot^{-1}a - \sec^{-1}b = \cos^{-1}\dfrac{a+\sqrt{b^2-1}}{b\sqrt{a^2+1}}.$

EXAMPLES.

3. From the formula $\cot 2x = \frac{\cot^2 x - 1}{2\cot x}$, prove

$$2\cot^{-1} a = \cot^{-1}\frac{a^2-1}{2a}.$$

4. From the formula $\cos 2x = 1 - 2\sin^2 x$, prove

$$2\sin^{-1} a = \cos^{-1}(1-2a^2).$$

5. From the formula $\sin 2x = 2\sin x\cos x$, prove

$$2\cos^{-1} a = \sin^{-1}(2a\sqrt{1-a^2}).$$

6. From the formula $\cos(x+y) = \cos x\cos y - \sin x\sin y$, prove

$$\cos^{-1} a + \cos^{-1} b = \cos^{-1}(ab - \sqrt{1-a^2}\sqrt{1-b^2}).$$

7. From the formula $\sin 3x = 3\sin x - 4\sin^3 x$, prove

$$3\sin^{-1} a = \sin^{-1}(3a - 4a^3).$$

Prove the following relations:

8. $\cot^{-1} a + \cot^{-1} b = \cot^{-1}\frac{ab-1}{a+b}.$

9. $2\cos^{-1} a = \cos^{-1}(2a^2 - 1).$

10. $\sin^{-1} a - \sin^{-1} b = \sin^{-1}(a\sqrt{1-b^2} - b\sqrt{1-a^2}).$

11. $3\tan^{-1} a = \tan^{-1}\frac{3a-a^3}{1-3a^2}.$

12. $\cot^{-1}(a-b) - \cot^{-1}(a+b) = \cot^{-1}\frac{a^2-b^2+1}{2b}.$

13. $\sin^{-1} a + \cos^{-1} b = \tan^{-1}\frac{ab+\sqrt{1-a^2}\sqrt{1-b^2}}{b\sqrt{1-a^2} - a\sqrt{1-b^2}}.$

14. $\sec^{-1} a - \csc^{-1} b = \cos^{-1}\frac{\sqrt{a^2-1}+\sqrt{b^2-1}}{ab}.$

15. $\tan^{-1} a + \cos^{-1}\frac{1}{a} = \sin^{-1}\frac{a+\sqrt{a^2-1}}{a\sqrt{a^2+1}}.$

16. $\tan^{-1}\frac{a}{a-1} - \tan^{-1}\frac{a+1}{a} = \tan^{-1}\frac{1}{2a^2}.$

17. $2\sin^{-1} a = \tan^{-1}\frac{2a\sqrt{1-a^2}}{1-2a^2}.$

18. $\tan^{-1} a + 2\tan^{-1} b = \tan^{-1}\frac{a(1-b^2)+2b}{1-b^2-2ab}.$

59. The following table expresses the value of each of the six principal functions of an angle in terms of the other five:

$\sin A$	……	$\sqrt{1-\cos^2 A}$	$\frac{\tan A}{\sqrt{1+\tan^2 A}}$	$\frac{1}{\sqrt{1+\cot^2 A}}$	$\frac{\sqrt{\sec^2 A-1}}{\sec A}$	$\frac{1}{\csc A}$
$\cos A$	$\sqrt{1-\sin^2 A}$	……	$\frac{1}{\sqrt{1+\tan^2 A}}$	$\frac{\cot A}{\sqrt{1+\cot^2 A}}$	$\frac{1}{\sec A}$	$\frac{\sqrt{\csc^2 A-1}}{\csc A}$
$\tan A$	$\frac{\sin A}{\sqrt{1-\sin^2 A}}$	$\frac{\sqrt{1-\cos^2 A}}{\cos A}$	……	$\frac{1}{\cot A}$	$\sqrt{\sec^2 A-1}$	$\frac{1}{\sqrt{\csc^2 A-1}}$
$\cot A$	$\frac{\sqrt{1-\sin^2 A}}{\sin A}$	$\frac{\cos A}{\sqrt{1-\cos^2 A}}$	$\frac{1}{\tan A}$	……	$\frac{1}{\sqrt{\sec^2 A-1}}$	$\sqrt{\csc^2 A-1}$
$\sec A$	$\frac{1}{\sqrt{1-\sin^2 A}}$	$\frac{1}{\cos A}$	$\sqrt{1+\tan^2 A}$	$\frac{\sqrt{1+\cot^2 A}}{\cot A}$	……	$\frac{\csc A}{\sqrt{\csc^2 A-1}}$
$\csc A$	$\frac{1}{\sin A}$	$\frac{1}{\sqrt{1-\cos^2 A}}$	$\frac{\sqrt{1+\tan^2 A}}{\tan A}$	$\sqrt{1+\cot^2 A}$	$\frac{\sec A}{\sqrt{\sec^2 A-1}}$	……

The reciprocal forms were proved in § 36.

The others may be derived by aid of §§ 36, 37, 38, 39, and 40, and are left as exercises for the student.

As an illustration, we will give a proof of the formula

$$\cos A = \frac{\sqrt{\csc^2 A - 1}}{\csc A}.$$

By § 39, $\cos A = \sqrt{1-\sin^2 A} = \sqrt{1 - \frac{1}{\csc^2 A}} = \frac{\sqrt{\csc^2 A - 1}}{\csc A}.$

They may also be conveniently proved by the method of § 6; thus, let it be required to prove the formula for each of the other functions in terms of the secant.

We have $$\sec A = \frac{\sec A}{1}.$$

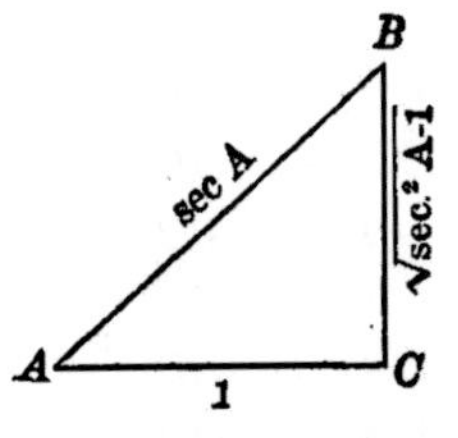

Since the secant is the ratio of the hypotenuse to the adjacent side, we take $AB = \sec A$, and $AC = 1$; whence, $BC = \sqrt{\overline{AB}^2 - \overline{AC}^2} = \sqrt{\sec^2 A - 1}.$

Then by definition,

$$\sin A = \frac{\sqrt{\sec^2 A - 1}}{\sec A}, \qquad \tan A = \sqrt{\sec^2 A - 1}, \qquad \csc A = \frac{\sec A}{\sqrt{\sec^2 A - 1}}.$$

$$\cos A = \frac{1}{\sec A}, \qquad \cot A = \frac{1}{\sqrt{\sec^2 A - 1}},$$

60. Line Values of the Functions.

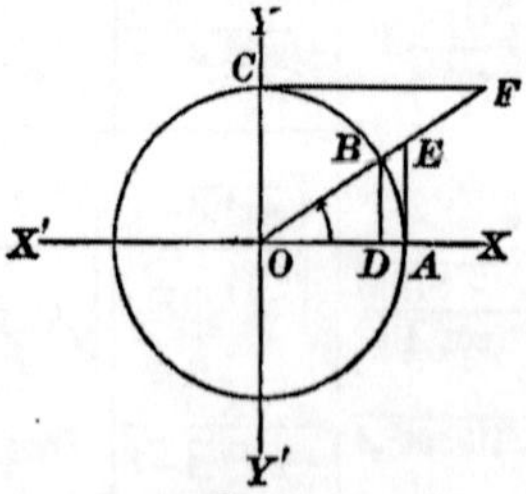

Fig. 1.

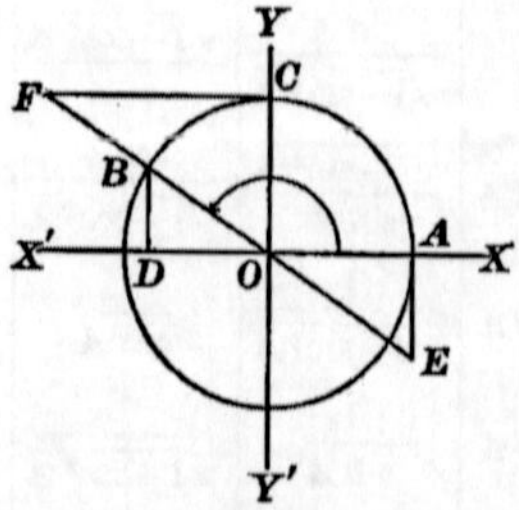

Fig. 2.

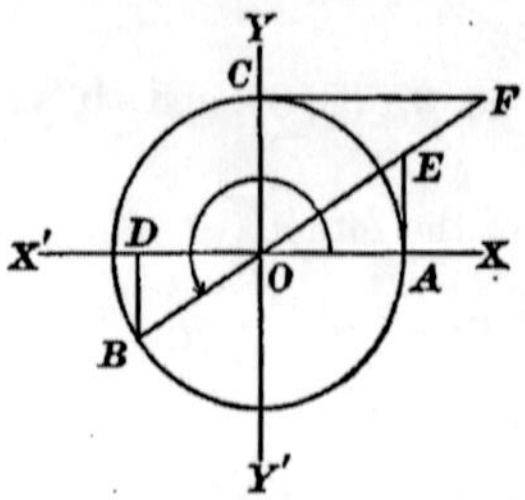

Fig. 3.

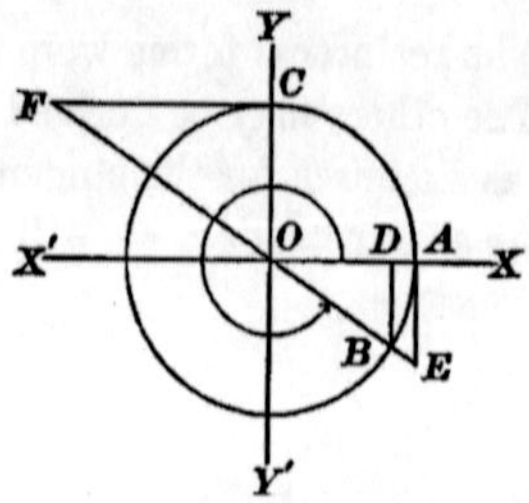

Fig. 4.

Let AOB be any angle. With O as a centre, and a radius equal to 1, describe the circle AB; draw BD and AE perpendicular to XX', and CF perpendicular to YY'. Then by § 17, the functions of AOB are:

	Sin.	*Cos.*	*Tan.*	*Cot.*	*Sec.*	*Csc.*
Fig. 1.	$\frac{BD}{OB}$	$\frac{OD}{OB}$	$\frac{BD}{OD}$	$\frac{OD}{BD}$	$\frac{OB}{OD}$	$\frac{OB}{BD}$
Fig. 2.	$\frac{BD}{OB}$	$-\frac{OD}{OB}$	$-\frac{BD}{OD}$	$-\frac{OD}{BD}$	$-\frac{OB}{OD}$	$\frac{OB}{BD}$
Fig. 3.	$-\frac{BD}{OB}$	$-\frac{OD}{OB}$	$\frac{BD}{OD}$	$\frac{OD}{BD}$	$-\frac{OB}{OD}$	$-\frac{OB}{BD}$
Fig. 4.	$-\frac{BD}{OB}$	$\frac{OD}{OB}$	$-\frac{BD}{OD}$	$-\frac{OD}{BD}$	$\frac{OB}{OD}$	$-\frac{OB}{BD}$

But since the right triangles OBD, OEA, and OCF are similar, and $OA = OC = 1$, we have

$$\frac{BD}{OD} = \frac{AE}{OA} = AE, \qquad \frac{OB}{OD} = \frac{OE}{OA} = OE,$$

$$\frac{OD}{BD} = \frac{CF}{OC} = CF, \qquad \frac{OB}{BD} = \frac{OF}{OC} = OF.$$

Whence, since $OB = 1$, the functions of AOB are:

	Sin.	*Cos.*	*Tan.*	*Cot.*	*Sec.*	*Csc.*
Fig. 1.	BD	OD	AE	CF	OE	OF
Fig. 2.	BD	$-OD$	$-AE$	$-CF$	$-OE$	OF
Fig. 3.	$-BD$	$-OD$	AE	CF	$-OE$	$-OF$
Fig. 4.	$-BD$	OD	$-AE$	$-CF$	OE	$-OF$

That is, *if the radius of the circle is* 1,

The *sine* is the perpendicular drawn to XX' from the intersection of the circle with the terminal line.

The *cosine* is the line drawn from the centre to the foot of the sine.

The *tangent* is that portion of the geometrical tangent to the circle at its intersection with OX included between OX and the terminal line, produced if necessary.

The *cotangent* is that portion of the geometrical tangent to the circle at its intersection with OY included between OY and the terminal line, produced if necessary.

The *secant* is that portion of the terminal line, or terminal line produced, included between the centre and the tangent.

The *cosecant* is that portion of the terminal line, or terminal line produced, included between the centre and the cotangent.

And with regard to algebraic signs,

Sines and tangents measured *above* XX' are *positive*, and *below*, *negative;* cosines and cotangents measured to the *right* of YY' are *positive*, and to the *left*, *negative;* secants and cosecants measured on the terminal line itself are *positive*, and on the terminal line *produced*, *negative.*

The above are called the *line values* of the trigonometric functions.

They simply *represent* the values of the functions when the radius is 1; that is, the *numerical value* of the sine of an angle is the same as the *number* which expresses the length of the perpendicular drawn to XX' from the intersection of the circle and terminal line.

61. *To trace the changes in the six principal trigonometric functions of an angle as the angle increases from 0° to 360°.*

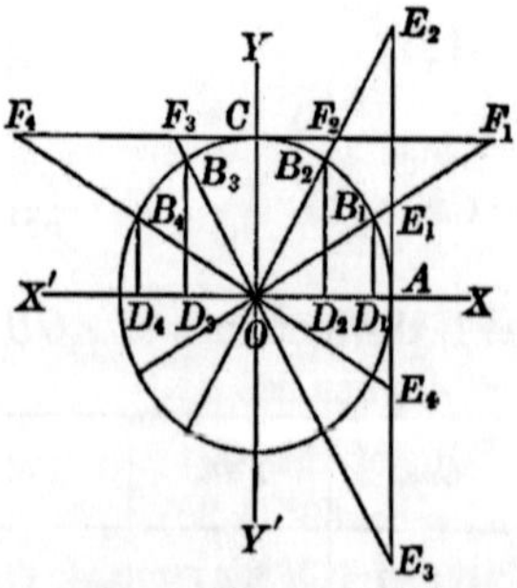

Let the terminal line start from the position OA, and revolve about the point O as a pivot, in a direction contrary to the motion of the hands of a clock.

Then since the sine of the angle commences with the value 0, and assumes in succession the values B_1D_1, B_2D_2, OC, B_3D_3, B_4D_4, etc. (§ 60), it is evident that, as the angle increases from 0° to 90°, the sine increases from 0 to 1; from 90° to 180°, it decreases from 1 to 0; from 180° to 270°, it decreases (algebraically) from 0 to -1; and from 270° to 360°, it increases from -1 to 0.

Since the cosine commences with the value OA, and assumes in succession the values OD_1, OD_2, 0, $-OD_3$, $-OD_4$, etc., from 0° to 90°, it decreases from 1 to 0; from 90° to 180°, it decreases from 0 to -1; from 180° to 270°, it increases from -1 to 0; and from 270° to 360°, it increases from 0 to 1.

Since the tangent commences with the value 0, and assumes in succession the values AE_1, AE_2, ∞, $-AE_3$, $-AE_4$, etc., from 0° to 90°, it increases from 0 to ∞; from 90° to 180°, it increases from $-\infty$ to 0; from 180° to 270°, it increases from 0 to ∞; and from 270° to 360°, it increases from $-\infty$ to 0.

Since the cotangent commences at ∞, and assumes in succession the values CF_1, CF_2, 0, $-CF_3$, $-CF_4$, etc., from 0° to 90°, it decreases from ∞ to 0; from 90° to 180°, it decreases from 0 to $-\infty$; from 180° to 270°, it decreases from ∞ to 0; and from 270° to 360°, it decreases from 0 to $-\infty$.

Since the secant commences with the value OA, and assumes in succession the values OE_1, OE_2, ∞, $-OE_3$, $-OE_4$, etc., from 0° to 90°, it increases from 1 to ∞; from 90° to 180°, it increases from $-\infty$ to -1; from 180° to 270°, it decreases from -1 to $-\infty$; and from 270° to 360°, it decreases from ∞ to 1.

Since the cosecant commences at ∞, and assumes in succession the values OF_1, OF_2, OC, OF_3, OF_4, etc., from 0° to 90°, it decreases from ∞ to 1; from 90° to 180°, it increases from 1 to ∞; from 180° to 270°, it increases from $-\infty$ to -1; and from 270° to 360°, it decreases from -1 to $-\infty$.

Note. Wherever the symbol ∞ occurs in the above discussion, it must be interpreted as explained in the Note to § 25.

62. Trigonometric Equations.

1. Find the value of A when $\cos A = \frac{1}{2}$.

We know that one value of A is 60° (§ 8).

And since $\cos(-60°) = \cos 60°$ (§ 29), another value of A is $-60°$.

Again, by the principle of § 21, any multiple of 360° may be added to, or subtracted from, an angle, without altering its functions.

Hence, other values of A are

$$360°+60°,\ 720°+60°,\ -360°+60°,\ 360°-60°,\ 720°-60°,\ -360°-60°, \text{ etc.}$$

It is evident from the above that the number of possible values of A is indefinitely great; and that each is in the form

$$n \times 360° + 60°, \text{ or } n \times 360° - 60°;$$

where n is 0, or any positive or negative integer.

Using the circular notation, we have

$$A = n \times 2\pi \pm \frac{\pi}{3} = 2n\pi \pm \frac{\pi}{3}.$$

2. Find the value of A when $\tan A = \frac{1}{3}\sqrt{3}$.

We know that one value of A is 30° (§ 8); another is 180°+30° (§ 27).

Adding to, and subtracting from, these angles multiples of 360°, other values of A are

$$360° + 30°,\ 540° + 30°,\ -360° + 30°,\ -180° + 30°, \text{ etc.}$$

It is evident from the above that all the values of A are given by the expression

$$n \times 180° + 30°;$$

where n is 0, or any positive or negative integer.

Or, $$A = n\pi + \frac{\pi}{6}.$$

3. Find the value of A when $\sin A = \frac{1}{2}\sqrt{2}$.

One value of A is 45° (§ 7).

And since $\sin(180° - 45°) = \sin 45°$ (§ 33), another value of A is $180° - 45°$.

Adding to, and subtracting from, these angles multiples of 360°, other values of A are

$$360° + 45°,\ 540° - 45°,\ -360° + 45°,\ -180° - 45°, \text{ etc.}$$

It is evident from the above that all the values of A are given by the expression

$$n \times 180° + (-1)^n 45°;$$

where n is 0, or any positive or negative integer.

Or, $$A = n\pi + (-1)^n \frac{\pi}{4}.$$

It is evident that, to find the value of A in any equation of the above forms, we find *any one of the values of A*, and substitute it for A in the following expressions:

If $\sin A$ is given, $n\pi + (-1)^n A$.

If $\cos A$ is given, $2n\pi \pm A$.

If $\tan A$ is given, $n\pi + A$.

The rule for equations giving the value of $\cot A$ is the same as for $\tan A$; for $\sec A$, the same as for $\cos A$; and for $\csc A$, the same as for $\sin A$.

EXAMPLES.

In each of the following find the value of A:

4. $\tan A = \sqrt{3}$.
6. $\sin A = \frac{1}{2}$.
8. $\cot A = -1$.
10. $\cot A = 0$.
5. $\cos A = -\frac{1}{2}\sqrt{3}$.
7. $\sec A = \sqrt{2}$.
9. $\csc A = -\frac{2}{3}\sqrt{3}$.
11. $\sec A = -1$.

63. 1. Solve the equation $\cos 2A = \cos A$.

By (**28**), $2\cos^2 A - 1 = \cos A$, or $2\cos^2 A - \cos A = 1$.

Solving this equation, $\cos A = \dfrac{1 \pm \sqrt{1+8}}{4} = \dfrac{1 \pm 3}{4} = 1$ or $-\dfrac{1}{2}$.

If $\cos A = 1$, one value of A is 0° (§ 22), and $A = 2n\pi$ (§ 62).

If $\cos A = -\dfrac{1}{2}$, one value of A is 120° (§ 27), and $A = 2n\pi \pm \dfrac{2\pi}{3}$.

2. Solve the equation $\tan 2x = 6 \tan x$.

By (**29**), $$\frac{2\tan x}{1 - \tan^2 x} = 6 \tan x. \qquad (A)$$

One solution is evidently $\tan x = 0$.

In this case, one value of x is 0°, and $x = n\pi$ (§ 62).

Dividing (A) by $2\tan x$, we have

$$\frac{1}{1-\tan^2 x}=3,\text{ or }1=3-3\tan^2 x,\text{ or }3\tan^2 x=2.$$

Whence, $\tan^2 x=\frac{2}{3}$, or $\tan x=\pm\sqrt{\frac{2}{3}}=\pm\frac{1}{3}\sqrt{6}$.

Therefore, $x=\tan^{-1}(\pm\frac{1}{3}\sqrt{6})=\pm\tan^{-1}(\frac{1}{3}\sqrt{6})$.

EXAMPLES.

In each of the following find the value of x:

3. $\sin x=\sin 2x$.
4. $\sin 2x+\cos x=0$.
5. $\cos x+\cos 3x=0$.
6. $\tan 3x+\tan x=0$.
7. $\cot 2x+\cot x=0$.
8. $\tan(45°-x)+\cot(45°-x)=4$.
9. $\tan 3x=5\tan x$.
10. $\cos x\cot x=1$.

64. Limiting Values of $\frac{\sin x}{x}$ and $\frac{\tan x}{x}$.

To find the limiting values of the fractions $\frac{\sin x}{x}$ *and* $\frac{\tan x}{x}$ *when* x *is indefinitely decreased.*

Note. We suppose x to be expressed in *circular measure* (§ 52).

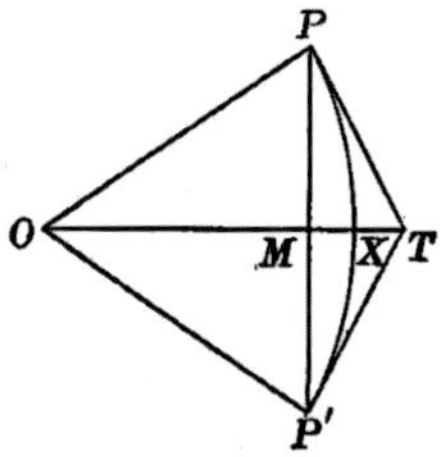

Let $OPXP'$ be a sector of a circle.

Draw PT and $P'T$ tangent to the arc at P and P', and join OT and PP'.

By Geometry, $PT=P'T$.

Then OT is perpendicular to PP' at its middle point M, and bisects the arc PP' at X.

Let $\angle XOP=\angle XOP'=x$.

By Geometry, arc $PP'>$ chord PP', and $<PTP'$.

Whence, arc $PX>PM$, and $<PT$.

Therefore, $\frac{\text{arc }PX}{OP}>\frac{PM}{OP}$, and $<\frac{PT}{OP}$.

Or by § 52, circ. meas. $x>\sin x$, and $<\tan x$.

Representing the circular measure of x by x simply, and dividing through by $\sin x$, we have

$$\frac{x}{\sin x} > 1, \text{ and } < \frac{\tan x}{\sin x} \text{ or } \frac{1}{\cos x}.$$

Whence, $$\frac{\sin x}{x} < 1, \text{ and } > \cos x.$$

But when x is indefinitely decreased, $\cos x$ approaches the limit 1 (§ 22).

Hence, $\frac{\sin x}{x}$ approaches the limit 1 when x is indefinitely decreased.

Again, $$\frac{\tan x}{x} = \frac{\sin x}{x \cos x} = \frac{\sin x}{x} \times \frac{1}{\cos x}.$$

But $\frac{\sin x}{x}$ and $\frac{1}{\cos x}$ approach the limit 1 when x is indefinitely decreased.

Hence, $\frac{\tan x}{x}$ approaches the limit 1 when x is indefinitely decreased.

V. LOGARITHMS.

65. Every positive number may be expressed, exactly or approximately, as a power of 10.

Thus, $100 = 10^2$; $13 = 10^{1.113943\ldots}$; etc.

When thus expressed, the corresponding exponent is called its **Logarithm to the Base 10.**

Thus, 2 is the logarithm of 100 to the base 10; a relation which is written $\log_{10} 100 = 2$, or simply $\log 100 = 2$.

66. Logarithms of numbers to the base 10 are called *Common Logarithms*, and, collectively, form the *Common System.*

They are the only ones used for numerical computations.

Any positive number, except unity, may be taken as the base of a system of logarithms; thus, if $a^x = m$, where a and m are positive numbers, then $x = \log_a m$.

Note. A negative number is not considered as having a logarithm.

67. We have by Algebra,

$$10^0 = 1, \qquad 10^{-1} = \frac{1}{10} = .1,$$

$$10^1 = 10, \qquad 10^{-2} = \frac{1}{10^2} = .01,$$

$$10^2 = 100, \qquad 10^{-3} = \frac{1}{10^3} = .001, \text{ etc.}$$

Whence by the definition of § 65,

$$\log 1 = 0, \qquad \log .1 = -1 = 9 - 10,$$

$$\log 10 = 1, \qquad \log .01 = -2 = 8 - 10,$$

$$\log 100 = 2, \qquad \log .001 = -3 = 7 - 10, \text{ etc.}$$

Note. The second form for log .1, log .01, etc., is preferable in practice. If no base is expressed, the base 10 is understood.

68. It is evident from § 67 that the logarithm of a number greater than 1 is positive, and the logarithm of a number between 0 and 1 negative.

69. If a number is not an exact power of 10, its common logarithm can only be expressed approximately.

The integral part of the logarithm is called the *characteristic,* and the decimal part the *mantissa.*

For example, $\log 13 = 1.113943.$

In this case, the characteristic is 1, and the mantissa .113943.

For reasons which will appear hereafter, only the mantissa of the logarithm is given in a table of logarithms of numbers; the characteristic must be found by aid of the rules of §§ 70 and 71.

70. It is evident from § 67 that the logarithm of a number between

1 and 10 is equal to 0 + a decimal;
10 and 100 is equal to 1 + a decimal;
100 and 1000 is equal to 2 + a decimal; etc.

Therefore, the characteristic of the logarithm of a number with *one* figure to the left of the decimal point, is 0; with *two* figures to the left of the decimal point, is 1; with *three* figures to the left of the decimal point, is 2; etc.

Hence, *the characteristic of the logarithm of a number greater than* 1 *is* 1 *less than the number of places to the left of the decimal point.*

For example, the characteristic of log 906328.51 is 5.

71. In like manner, the logarithm of a number between

1 and .1 is equal to 9 + a decimal − 10;
.1 and .01 is equal to 8 + a decimal − 10;
.01 and .001 is equal to 7 + a decimal − 10; etc.

Therefore, the characteristic of the logarithm of a decimal with *no* ciphers between its decimal point and first significant figure, is 9, with − 10 after the mantissa; of a decimal with *one* cipher between its point and first significant figure is 8, with − 10 after the mantissa; of a decimal with *two* ciphers between its point and first significant figure is 7, with − 10 after the mantissa; etc.

Hence, *to find the characteristic of the logarithm of a number less than* 1, *subtract the number of ciphers between the decimal point and first significant figure from* 9, *writing* − 10 *after the mantissa.*

For example, the characteristic of log .007023 is 7, with − 10 written after the mantissa.

Note. Some writers combine the two portions of the characteristic, and write the result as a *negative characteristic* before the mantissa.

Thus, instead of 7.603658 − 10, the student will frequently find $\bar{3}.603658$, a minus sign being written over the characteristic to denote that it alone is negative, the mantissa being always positive.

PROPERTIES OF LOGARITHMS.

72. *In any system, the logarithm of* 1 *is* 0.

For by Algebra, $a^0 = 1$; whence by § 66, $\log_a 1 = 0$.

73. *In any system, the logarithm of the base is* 1.

For $a^1 = a$; whence, $\log_a a = 1$.

74. *In any system whose base is greater than* 1, *the logarithm of* 0 *is* $-\infty$.

For if a is greater than 1, $a^{-\infty} = \frac{1}{a^\infty} = \frac{1}{\infty} = 0$.

Whence by § 66, $\log_a 0 = -\infty$.

Note. No literal meaning can be attached to such a result as $\log_a 0 = -\infty$; it must be interpreted as follows:

If, in any system whose base is greater than unity, a number approaches the limit 0, its logarithm is negative, and increases without limit in absolute value.

75. *In any system, the logarithm of a product is equal to the sum of the logarithms of its factors.*

Assume the equations

$$\left.\begin{array}{l} a^x = m \\ a^y = n \end{array}\right\}; \text{ whence by § 66, } \left\{\begin{array}{l} x = \log_a m, \\ y = \log_a n. \end{array}\right.$$

Multiplying the assumed equations,

$$a^x \times a^y = mn, \text{ or } a^{x+y} = mn.$$

Whence, $\log_a mn = x + y = \log_a m + \log_a n$.

In like manner, the theorem may be proved for the product of three or more factors.

76. By aid of § 75, the logarithm of a composite number may be found when the logarithms of its factors are known.

1. Given $\log 2 = .3010$ and $\log 3 = .4771$; find $\log 72$.

$$\begin{aligned} \log 72 &= \log(2 \times 2 \times 2 \times 3 \times 3) \\ &= \log 2 + \log 2 + \log 2 + \log 3 + \log 3 \text{ (§ 75)} \\ &= 3 \times \log 2 + 2 \times \log 3 = .9030 + .9542 = 1.8572. \end{aligned}$$

EXAMPLES.

Given $\log 2 = .3010$, $\log 3 = .4771$, $\log 5 = .6990$, $\log 7 = .8451$, find:

2. $\log 35$.	**6.** $\log 126$.	**10.** $\log 324$.	**14.** $\log 2625$.
3. $\log 50$.	**7.** $\log 196$.	**11.** $\log 378$.	**15.** $\log 6048$.
4. $\log 42$.	**8.** $\log 245$.	**12.** $\log 875$.	**16.** $\log 12005$.
5. $\log 75$.	**9.** $\log 210$.	**13.** $\log 686$.	**17.** $\log 15876$.

77. *In any system, the logarithm of a fraction is equal to the logarithm of the numerator minus the logarithm of the denominator.*

Assume the equations

$$\left.\begin{matrix} a^x = m \\ a^y = n \end{matrix}\right\}; \text{ whence, } \left\{\begin{matrix} x = \log_a m, \\ y = \log_a n. \end{matrix}\right.$$

Dividing the assumed equations,

$$\frac{a^x}{a^y} = \frac{m}{n}, \text{ or } a^{x-y} = \frac{m}{n}.$$

Whence,
$$\log_a \frac{m}{n} = x - y = \log_a m - \log_a n.$$

78. 1. Given $\log 2 = .3010$; find $\log 5$.

$$\log 5 = \log \frac{10}{2} = \log 10 - \log 2 \ (\S\ 77) = 1 - .3010 = .6990.$$

EXAMPLES.

Given $\log 2 = .3010$, $\log 3 = .4771$, $\log 7 = .8451$, find:

2. $\log \frac{10}{3}$.	**5.** $\log 14\frac{2}{7}$.	**8.** $\log \frac{48}{25}$.	**11.** $\log 28\frac{4}{5}$.
3. $\log \frac{7}{4}$.	**6.** $\log \frac{49}{27}$.	**9.** $\log 6\frac{2}{9}$.	**12.** $\log \frac{200}{9}$.
4. $\log 45$.	**7.** $\log 225$.	**10.** $\log 135$.	**13.** $\log 110\frac{1}{4}$.

79. *In any system, the logarithm of any power of a quantity is equal to the logarithm of the quantity multiplied by the exponent of the power.*

Assume the equation $a^x = m$; whence, $x = \log_a m$.

Raising both members of the assumed equation to the pth power,

$$a^{px} = m^p; \text{ whence, } \log_a m^p = px = p \log_a m.$$

80. *In any system, the logarithm of any root of a quantity is equal to the logarithm of the quantity divided by the index of the root.*

For, $$\log_a \sqrt[r]{m} = \log_a (m^{\frac{1}{r}}) = \frac{1}{r} \log_a m \text{ (§ 79).}$$

81. 1. Given $\log 2 = .3010$; find $\log 2^{\frac{5}{3}}$.

$$\log 2^{\frac{5}{3}} = \frac{5}{3} \times \log 2 = \frac{5}{3} \times .3010 = .5017.$$

Note. To multiply a logarithm by a fraction, multiply first by the numerator, and divide the result by the denominator.

2. Given $\log 3 = .4771$; find $\log \sqrt[8]{3}$.

$$\log \sqrt[8]{3} = \frac{\log 3}{8} = \frac{.4771}{8} = .0596.$$

EXAMPLES.

Given $\log 2 = .3010$, $\log 3 = .4771$, $\log 7 = .8451$, find:

3. $\log 3^7$.
4. $\log 5^{\frac{5}{2}}$.
5. $\log 7^{\frac{3}{4}}$.
6. $\log 28^6$.
7. $\log 18^{\frac{5}{6}}$.
8. $\log 96^{\frac{2}{7}}$.
9. $\log \sqrt[7]{2}$.
10. $\log \sqrt[6]{5}$.
11. $\log \sqrt[9]{7}$.
12. $\log \sqrt[5]{525}$.
13. $\log \sqrt[4]{294}$.
14. $\log \sqrt[8]{216}$.

15. Find $\log (2^{\frac{1}{3}} \times 3^{\frac{5}{4}})$.

By § 75, $\log (2^{\frac{1}{3}} \times 3^{\frac{5}{4}}) = \log 2^{\frac{1}{3}} + \log 3^{\frac{5}{4}} = \frac{1}{3} \log 2 + \frac{5}{4} \log 3$
$= .1003 + .5964 = .6967.$

Find the values of the following:

16. $\log \sqrt[11]{\frac{7}{2}}$.
17. $\log \left(\frac{7}{5}\right)^{\frac{1}{4}}$.
18. $\log (2^{\frac{2}{3}} \times 10^{\frac{1}{2}})$.
19. $\log 7\sqrt[10]{2}$.
20. $\log \frac{\sqrt[4]{5}}{\sqrt[5]{3}}$.
21. $\log \frac{3^{\frac{4}{3}}}{7^{\frac{1}{4}}}$.
22. $\log \frac{3^{\frac{5}{3}}}{\sqrt{24}}$.
23. $\log \frac{\sqrt[3]{63}}{5^{\frac{4}{5}}}$.

82. *To prove the relation*

$$\log_b m = \frac{\log_a m}{\log_a b}.$$

Assume the equations

$$\left.\begin{aligned} a^x &= m \\ b^y &= m \end{aligned}\right\}; \text{ whence, } \left\{\begin{aligned} x &= \log_a m, \\ y &= \log_b m. \end{aligned}\right.$$

From the assumed equations, $a^x = b^y$.

Taking the yth root of both members, $a^{\frac{x}{y}} = b$.

Therefore, $$\log_a b = \frac{x}{y}, \text{ or } y = \frac{x}{\log_a b}.$$

That is, $$\log_b m = \frac{\log_a m}{\log_a b}.$$

83. *To prove the relation*

$$\log_b a \times \log_a b = 1.$$

Putting $m = a$ in the result of § 82, we have

$$\log_b a = \frac{\log_a a}{\log_a b} = \frac{1}{\log_a b} \text{ (§ 73).}$$

Whence, $$\log_b a \times \log_a b = 1.$$

84. *In the common system, the mantissæ of the logarithms of numbers having the same sequence of figures are equal.*

Suppose, for example, that $\log 3.053 = .484727$.

Then, $$\log 305.3 = \log (100 \times 3.053) = \log 100 + \log 3.053$$
$$= 2 + .484727 = 2.484727;$$
$$\log .03053 = \log (.01 \times 3.053) = \log .01 + \log 3.053$$
$$= 8 - 10 + .484727 = 8.484727 - 10; \text{ etc.}$$

It is evident from the above that, if a number be multiplied or divided by any integral power of 10, producing another number with the same sequence of figures, the mantissæ of their logarithms will be equal.

The reason will now be seen for the statement made in § 69, that only the mantissæ are given in a table of logarithms of numbers.

For, to find the logarithm of any number, we have only to take from the table the mantissa corresponding to its sequence of figures, and the characteristic may then be prefixed in accordance with the rules of §§ 70 or 71.

Thus, if $\log 3.053 = .484727$, then

$$\log 30.53 = 1.484727, \qquad \log .3053 = 9.484727 - 10,$$
$$\log 305.3 = 2.484727, \qquad \log .03053 = 8.484727 - 10,$$
$$\log 3053. = 3.484727, \qquad \log .003053 = 7.484727 - 10, \text{ etc.}$$

This property is only enjoyed by the common system of logarithms, and constitutes its superiority over others for the purposes of numerical computation.

85. 1. Given $\log 2 = .3010$, $\log 3 = .4771$; find $\log .00432$.

We have, $\log 432 = \log (2^4 \times 3^3) = 4 \log 2 + 3 \log 3 = 2.6353$.

Then by § 84, the *mantissa* of the result is .6353.

Whence by § 71, $\log .00432 = 7.6353 - 10$.

EXAMPLES.

Given $\log 2 = .3010$, $\log 3 = .4771$, $\log 7 = .8451$, find:

2. $\log 3.6$. **6.** $\log .00343$. **10.** $\log .1944$.

3. $\log 11.2$. **7.** $\log 2880$. **11.** $\log 202.5$.

4. $\log .84$. **8.** $\log .0392$. **12.** $\log \sqrt[7]{6.4}$.

5. $\log .098$. **9.** $\log .000405$. **13.** $\log (14.7)^{\frac{2}{3}}$.

USE OF THE TABLE OF LOGARITHMS OF NUMBERS.

(For directions as to the use of the Table of Logarithms of Numbers, see pages iii to v of the Introduction to the Author's Six Place Logarithmic Tables.)

EXAMPLES.

86. Find the logarithms of the following numbers:

1. .053. **5.** 336.908. **9.** .001030746.

2. 51.8. **6.** .000602851. **10.** .00000876092.

3. .2956. **7.** 65000.63. **11.** 730407.8.

4. 1.0274. **8.** 9122.55. **12.** .0000436927.

Find the numbers corresponding to the following logarithms:

13. 1.880814. **17.** 8.044891 – 10. **21.** 3.990191.

14. 9.470410 – 10. **18.** 2.270293. **22.** 5.670180.

15. 0.820204. **19.** 7.350064 – 10. **23.** 6.535003 – 10.

16. 4.745126. **20.** 5.000027 – 10. **24.** 4.115658 – 10.

APPLICATIONS.

87. The approximate value of an arithmetical quantity, in which the operations indicated involve only multiplication, division, involution, or evolution, may be conveniently found by logarithms.

The utility of the process consists in the fact that addition takes the place of multiplication, subtraction of division, multiplication of involution, and division of evolution.

Note. In computations with six-place logarithms, the results cannot usually be depended upon to more than *six* significant figures.

88. 1. Find the value of $.0631 \times 7.208 \times .51272$.

By § 75, $\log(.0631 \times 7.208 \times .51272) = \log .0631 + \log 7.208 + \log .51272$.

$$\begin{aligned} \log .0631 &= 8.800029 - 10 \\ \log 7.208 &= 0.857815 \\ \log .51272 &= 9.709880 - 10 \end{aligned}$$

Adding, $\log$ of result $= 19.367724 - 20 = 9.367724 - 10$. (See Note 1.)

Number corresponding to $9.367724 - 10 = .233197$.

Note 1. If the sum is a negative logarithm, it should be written in such a form that the negative portion of the characteristic may be -10.

Thus, $19.367724 - 20$ is written in the form $9.367724 - 10$.

2. Find the value of $\dfrac{336.852}{7980.04}$.

By § 77, $$\log \frac{336.852}{7980.04} = \log 336.852 - \log 7980.04.$$

$$\begin{aligned} \log 336.852 &= 12.527439 - 10 \quad \text{(See Note 2.)} \\ \log 7980.04 &= 3.902005 \end{aligned}$$

Subtracting, $\log$ of result $= 8.625434 - 10$

Number corresponding $= .0422118$.

Note 2. To subtract a greater logarithm from a less, or to subtract a negative logarithm from a positive, increase the characteristic of the minuend by 10, writing -10 after the mantissa to compensate.

Thus, to subtract 3.902005 from 2.527439, write the minuend in the form $12.527439 - 10$; subtracting 3.902005 from this, the result is $8.625434 - 10$.

3. Find the value of $(.0980937)^5$.

By § 79, $$\log(.0980937)^5 = 5 \times \log .0980937.$$

$$\log .0980937 = 8.991641 - 10$$
$$\times 5$$
$$44.958205 - 50 = 4.958205 - 10. \quad \text{(See Note 1.)}$$

Number corresponding $= .0000090825$.

4. Find the value of $\sqrt[3]{.035063}$.

By § 80, $$\log \sqrt[3]{.035063} = \tfrac{1}{3} \log .035063.$$

$$\log .035063 = 8.544849 - 10$$
$$3)\underline{28.544849 - 30} \quad \text{(See Note 3.)}$$
$$9.514950 - 10$$

Number corresponding $= .327303$.

Note 3. To divide a negative logarithm, write it in such a form that the negative portion of the characteristic may be exactly divisible by the divisor, with -10 as the quotient.

Thus, to divide $8.544849-10$ by 3, we write the logarithm in the form $28.544849-30$; dividing this by 3, the quotient is $9.514950-10$.

89. Arithmetical Complement.

The *Arithmetical Complement* of the logarithm of a number, or, briefly, the *Cologarithm* of the number, is the logarithm of the reciprocal of that number.

Thus,
$$\text{colog}\,409=\log\frac{1}{409}=\log 1-\log 409.$$
$$\begin{aligned}\log 1&=10.\qquad\quad -10\ \text{(Note 2, § 88.)}\\ \log 409&=\underline{2.611723\qquad\quad}\end{aligned}$$
Then,
$$\text{colog}\,409=7.388277-10.$$
Again,
$$\text{colog}\,.067=\log\frac{1}{.067}=\log 1-\log .067.$$
$$\begin{aligned}\log 1&=10.\qquad\quad -10\\ \log .067&=\underline{8.826075-10}\end{aligned}$$
Then,
$$\text{colog}\,.067=1.173925.$$

It follows from the above that *the cologarithm of a number may be found by subtracting its logarithm from* $10-10$.

Note. The cologarithm may be obtained by subtracting the last *significant* figure of the logarithm from 10, and each of the others from 9, -10 being written after the result in the case of a positive logarithm.

90. *Example.* Find the value of $\dfrac{51.384}{8.709\times .0946}$

$$\log\frac{51.384}{8.709\times .0946}=\log\left(51.384\times\frac{1}{8.709}\times\frac{1}{.0946}\right)$$
$$=\log 51.384+\log\frac{1}{8.709}+\log\frac{1}{.0946}$$
$$=\log 51.384+\text{colog}\,8.709+\text{colog}\,.0946.$$

$$\begin{aligned}\log 51.384&=1.710828\\ \text{colog}\,8.709&=9.060032-10\\ \text{colog}\,.0946&=\underline{1.024109}\\ &1.794969=\log 62.369.\end{aligned}$$

It is evident from the above example that the logarithm of a fraction whose terms are composed of factors may be found by the following rule:

Add together the logarithms of the factors of the numerator, and the cologarithms of the factors of the denominator.

Note. The value of the above fraction may be found without using cologarithms, by the following formula:

$$\log\frac{51.384}{8.709 \times .0946} = \log 51.384 - \log(8.709 \times .0946)$$
$$= \log 51.384 - (\log 8.709 + \log .0946).$$

The advantage in the use of cologarithms is that the written work of computation is exhibited in a more compact form.

EXAMPLES.

Note. A *negative* quantity has no common logarithm (§ 66, Note).

If such quantities occur in computation, they should be treated as if they were positive, and the *sign* of the result determined irrespective of the logarithmic work.

Thus, in Ex. 2, § 91, the value of $84.759 \times (-2280.76)$ is obtained by finding the value of 84.759×2280.76, and putting a negative sign before the result. See also Ex. 29.

91. Find by logarithms the values of the following:

1. $3.1425 \times 603.93.$

2. $84.759 \times (-2280.76).$

3. $(-4.39182) \times (-.0703968).$

4. $.936537 \times .00117854.$

5. $\dfrac{4867.2}{765.16}.$ **6.** $\dfrac{1.05478}{34.9564}.$ **7.** $\dfrac{2.7085}{.0868097}.$ **8.** $\dfrac{-.000680239}{.00512643}.$

9. $\dfrac{3.89612 \times .6946}{4694.9 \times .00454}.$

10. $\dfrac{715 \times (-.024158)}{(-.5157) \times 1420.63}.$

11. $\dfrac{(-.870284) \times 3.73}{(-.06585) \times (-42.317)}.$

12. $\dfrac{.082136 \times (-73.39)}{.838 \times 2808.72}.$

13. $(7.7954)^4.$

14. $(.83287)^7.$

15. $(-25.1437)^3.$

16. $(.01)^{\frac{3}{4}}.$

17. $(-964.38)^{\frac{4}{5}}.$

18. $(.0951293)^{\frac{5}{2}}.$

19. $(.000105936)^{\frac{5}{3}}.$

20. $\sqrt{5}.$

21. $\sqrt[5]{2}.$

22. $\sqrt[9]{-6}.$

23. $\sqrt[8]{100}.$

24. $\sqrt[4]{.19946}.$

25. $\sqrt[6]{.0725628}.$

26. $\sqrt[3]{.002613874}.$

27. $\sqrt[7]{-.000951735}.$

28. Find the value of $\dfrac{2\sqrt[3]{5}}{3^{\frac{5}{6}}}.$

By § 90,

$$\log\frac{2\sqrt[3]{5}}{3^{\frac{5}{6}}} = \log 2 + \log\sqrt[3]{5} + \text{colog}\, 3^{\frac{5}{6}} = \log 2 + \tfrac{1}{3}\log 5 + \tfrac{5}{6}\,\text{colog}\, 3.$$

$$
\begin{array}{lr}
 & \log 2 = .301030 \\
\log 5 = .698970; \quad \text{divide by } 3 = & .232990 \\
\text{colog}\, 3 = 9.522879 - 10; \quad \text{multiply by } \tfrac{5}{6} = & \underline{9.602399 - 10} \\
 & .136419 = \log 1.36905.
\end{array}
$$

29. Find the value of $\sqrt[3]{\dfrac{-.032956}{7.96183}}$.

$$\log\sqrt[3]{\frac{.032956}{7.96183}}=\tfrac{1}{3}\log\frac{.032956}{7.96183}=\tfrac{1}{3}(\log .032956-\log 7.96183).$$

$$\begin{array}{r}\log .032956 = 8.517934-10\\ \log 7.96183 = 0.901013\\ \hline 3)\underline{27.616921-30}\\ 9.205640-10=\log .160561.\end{array}$$

Result, $-.160561$.

Find the values of the following:

30. $4^{\frac{4}{3}}\times 7^{\frac{2}{5}}$.

31. $\dfrac{3^{\frac{5}{7}}}{8^{\frac{2}{9}}}$.

32. $\sqrt[10]{\dfrac{79}{46}}$.

33. $\dfrac{(.001)^{\frac{3}{2}}}{\sqrt[5]{7}}$.

34. $\dfrac{\sqrt{.08}}{(-10)^{\frac{8}{5}}}$.

35. $\left(-\dfrac{4400}{6927.7}\right)^{\frac{4}{5}}$.

36. $\sqrt{\dfrac{276.85}{940}}$.

37. $\dfrac{5^{\frac{7}{4}}}{\sqrt[3]{-.1}}$.

38. $\dfrac{-\sqrt[4]{1000}}{(-.6)^{\frac{4}{3}}}$.

39. $\sqrt[6]{\dfrac{3}{5}}\div\sqrt[5]{\dfrac{7}{8}}$.

40. $\sqrt[3]{3}\times\sqrt[5]{5}\times\sqrt[7]{7}$.

41. $\left(\dfrac{76.1\times .05929}{1.3073}\right)^{\frac{3}{4}}$.

42. $\sqrt[3]{-\dfrac{75.438}{31.4\times .4146}}$.

43. $\dfrac{\sqrt[4]{.000965782}}{\sqrt[3]{.00497836}}$.

44. $\dfrac{-(.256929)^{\frac{6}{5}}}{(-.834574)^{\frac{7}{3}}}$.

45. $(25.4673)^{10}\times(-.052)^{12}$.

46. $\sqrt[8]{5106.526\times .000031093}$.

47. $(837.48\times .00943246)^{\frac{2}{7}}$.

48. $(4.867184)^{\frac{7}{2}}\times(.175437)^{\frac{1}{5}}$

49. $\dfrac{\sqrt{3.9285}\times\sqrt[4]{65.4775}}{\sqrt[6]{721.329}}$.

50. $\dfrac{(.5732)^{\frac{3}{5}}}{8693.84\times\sqrt[4]{.033074}}$.

51. $\dfrac{(-.00019162)^{\frac{2}{3}}\times\sqrt{68.18}}{-.2755653}$.

52. $\dfrac{\sqrt[4]{.052866}}{\sqrt{.374}\times\sqrt[9]{.007835912}}$.

EXPONENTIAL EQUATIONS.

92. An **Exponential Equation** is an equation of the form $a^x=b$.

To solve an equation of this form, take the logarithms of both members.

1. Given $31^x=23$; find the value of x.

Taking the logarithms of both members,

$$\log(31^x)=\log 23.$$

Whence by § 79, $x\log 31=\log 23.$

Then, $$x=\frac{\log 23}{\log 31}=\frac{1.361728}{1.491362}=.9130+.$$

2. Given $.2^x = 3$; find the value of x.

Taking the logarithms of both members,

$$x \log .2 = \log 3.$$

Whence, $x = \dfrac{\log 3}{\log .2} = \dfrac{.477121}{9.301030 - 10} = \dfrac{.477121}{-.698970} = -.6826+.$

EXAMPLES.

Solve the following equations:

3. $332.9^x = 5.178.$ **5.** $.0158^x = .0082958.$ **7.** $a^x = b^{2x}c^5.$

4. $.4162^x = 6.724.$ **6.** $5.3364^x = .744.$ **8.** $m^2 a^{\frac{3}{x}} = n^4.$

9. $6^{2x-3} = .0277778.$ **10.** $.7^{x^2+4x} = .16807.$

93. 1. Find the logarithm of .3 to the base 7.

By § 82, $\log_7 .3 = \dfrac{\log_{10} .3}{\log_{10} 7} = \dfrac{9.477121 - 10}{.845098} = \dfrac{-.522879}{.845098} = -.6187+.$

EXAMPLES.

Find the values of the following:

2. $\log_2 13.$ **4.** $\log_{.74} 6.2.$ **6.** $\log_{9.1} .362.$

3. $\log_5 .9.$ **5.** $\log_{.48} .087.$ **7.** $\log_{65} 4.3.$

Examples like the above may be solved by inspection if the number can be expressed as an exact power of the base.

8. Find the logarithm of 128 to the base 16.

Let $\log_{16} 128 = x$; then by § 66, $16^x = 128.$

That is, $(2^4)^x = 2^7$, or $2^{4x} = 2^7.$

Whence by inspection, $4x = 7$; and $x = \log_{16} 128 = \dfrac{7}{4}.$

9. Find the logarithm of 81 to the base 3.

10. Find the logarithm of 32 to the base 8.

11. Find the logarithm of $\frac{1}{3}$ to the base 27.

12. Find the logarithm of $\frac{1}{64}$ to the base $\frac{1}{32}$.

EXAMPLES IN THE USE OF TRIGONOMETRIC TABLES.

(For directions, see pages v to xi of the Introduction to the Author's Six Place Logarithmic Tables.)

94. Table of Logarithmic Sines, Cosines, etc.

Find the values of the following:

1. log sin 12° 48′ 52″.
2. log tan 67° 13′ 27″.
3. log cos 31° 5′ 43″.
4. log cot 53° 42′ 9″.
5. log cos 79° 54′ 35″.
6. log tan 8° 17′ 21″.
7. log cot 26° 30′ 14″.
8. log sec 45° 26′ 38″.
9. log csc 84° 9′ 56″.

Find the angles corresponding in the following:

10. log sin = 9.934232 – 10.
11. log cos = 9.923569 – 10.
12. log tan = 0.806571.
13. log cot = 0.282956.
14. log tan = 9.184367 – 10.
15. log cot = 9.404692 – 10.
16. log sec = 0.188783.
17. log csc = 0.400314.

95. Table of Natural Sines, Cosines, etc.

Find the values of the following:

1. sin 43° 17′ 35″.
2. cot 75° 50′ 19″.
3. cos 86° 21′ 46″.
4. tan 34° 48′ 23″.
5. sin 67° 9′ 54″.
6. cos 29° 35′ 8″.

Find the angles corresponding in the following:

7. tan = 1.2622.
8. cos = .96376.
9. sin = .91527.
10. cot = 1.7927

96. Auxiliary Table for Small Angles.

Find the values of the following:

1. log sin 1° 14′ 53″.
2. log tan 3° 42′ 8″.
3. log cot 2° 26′ 35″.

Find the angles corresponding in the following:

4. log sin = 8.233459 – 10.
5. log tan = 7.859872 – 10.
6. log cot = 1.546267.

VI. SOLUTION OF RIGHT TRIANGLES.

97. The *elements* of a triangle are its three sides and its three angles. We know by Geometry that a triangle is, in general, completely determined when three of its elements are known, provided one of them is a side.

The *solution* of a triangle is the process of computing the unknown from the given elements.

98. To solve a *right triangle*, two elements must be given in addition to the right angle, one of which must be a side.

The various cases which can occur may all be solved by aid of the following formulæ:

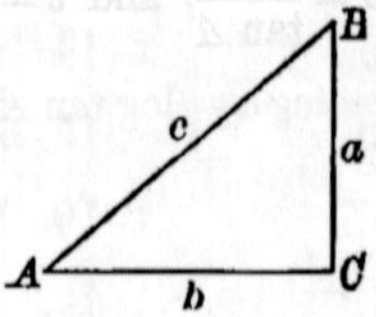

$$\sin A = \frac{a}{c}. \qquad \cos A = \frac{b}{c}. \qquad \tan A = \frac{a}{b}.$$

$$\sin B = \frac{b}{c}. \qquad \cos B = \frac{a}{c}. \qquad \tan B = \frac{b}{a}.$$

99. CASE I. *When the given elements are a side and an angle.*

The proper formula for computing either of the remaining sides may be found by the following rule:

Take that function of the angle which involves the given side and the required side.

1. Given $c = 68$, $B = 21° 42' 39''$; find a and b.

In this case the formulæ to be used are

$$\cos B = \frac{a}{c}, \text{ and } \sin B = \frac{b}{c}.$$

Whence, $\quad a = c \cos B, \text{ and } b = c \sin B.$ (A)

Solution by Natural Functions.

$$a = 68 \times \cos 21° 42' 39'' = 68 \times .92906 = 63.176.$$

$$b = 68 \times \sin 21° 42' 39'' = 68 \times .36993 = 25.155.$$

Solution by Logarithms.

Taking the logarithms of both members, in formulæ (A),

$$\log a = \log c + \log \cos B, \text{ and } \log b = \log c + \log \sin B.$$

$$\begin{array}{rl} \log c = & 1.832509 \\ \log \cos B = & 9.968045 - 10 \\ \hline \log a = & 1.800554 \\ a = & 63.1762. \end{array} \qquad \begin{array}{rl} \log c = & 1.832509 \\ \log \sin B = & 9.568111 - 10 \\ \hline \log b = & 1.400620 \\ b = & 25.1547. \end{array}$$

2. Given $a = .235867$, $A = 67° 9' 23''$; find b and c.

In this case, $\tan A = \frac{a}{b}$, and $\sin A = \frac{a}{c}$.

Whence, $b = \frac{a}{\tan A}$, and $c = \frac{a}{\sin A}$.

By logarithms, $\log b = \log a - \log \tan A$, and $\log c = \log a - \log \sin A$.

$$\begin{array}{rl} \log a = & 9.372667 - 10 \\ \log \tan A = & 0.375452 \\ \hline \log b = & 8.997215 - 10 \\ b = & .0993607. \end{array} \qquad \begin{array}{rl} \log a = & 9.372667 - 10 \\ \log \sin A = & 9.964527 - 10 \\ \hline \log c = & 9.408140 - 10 \\ c = & .255941. \end{array}$$

100. Case II. *When both the given elements are sides.*

First calculate one of the angles by aid of either formula involving the given elements, and then compute the remaining side by the rule of Case I.

Example. Given $b = .15124$, $c = .30807$; find A and a.

We first find A by the formula $\cos A = \frac{b}{c}$, and then find a by the formula $\sin A = \frac{a}{c}$, or $a = c \sin A$.

By logarithms, $\log \cos A = \log b - \log c$, and $\log a = \log c + \log \sin A$.

$$\begin{array}{rl} \log b = & 9.179667 - 10 \\ \log c = & 9.488650 - 10 \\ \hline \log \cos A = & 9.691017 - 10 \\ A = & 60° 35' 54.4''. \end{array} \qquad \begin{array}{rl} \log c = & 9.488650 - 10 \\ \log \sin A = & 9.940118 - 10 \\ \hline \log a = & 9.428768 - 10 \\ a = & .268391. \end{array}$$

101. In the Trigonometric solution of any example under Case II., it is necessary to first find one of the angles, and the remaining side may then be calculated.

It is possible, however, to compute the third side directly, without first finding the angle, by Geometry.

Thus, in the example of § 100, we have, by Geometry, $a^2 + b^2 = c^2$.

Whence, $$a = \sqrt{c^2 - b^2} = \sqrt{(c+b)(c-b)}.$$

By logarithms, $\log a = \frac{1}{2}[\log(c+b) + \log(c-b)]$.

$$c + b = .45931;\ \log = 9.662106 - 10$$
$$c - b = .15683;\ \log = 9.195429 - 10$$
$$2)\overline{18.857535 - 20}$$
$$\log a = 9.428768 - 10$$
$$a = .268391, \text{ as before.}$$

If the given sides are a and b, the expression for c is $\sqrt{a^2 + b^2}$, which is not adapted to logarithmic computation.

In such a case, it is usually shorter to proceed as in § 100.

EXAMPLES.

Note. In those examples of the following set in which the given sides are numbers of not more than three significant figures, and the operations indicated involve only multiplication, it is usually shorter to employ Natural Functions.

In such a case, the results cannot be depended upon to more than *five* significant figures; while in the solutions by logarithms, they can be depended upon to *six* significant figures.

102. Solve the following right triangles:

1. Given $A = 15°$, $c = 7$.
2. Given $B = 67°$, $a = 5$.
3. Given $B = 50°$, $b = 20$.
4. Given $a = .35$, $c = .62$.
5. Given $a = 273$, $b = 418$.
6. Given $A = 38°$, $a = 8.09$.
7. Given $B = 75°$, $c = .014$.
8. Given $b = 58.6$, $c = 76.3$.
9. Given $A = 9°$, $b = 937$.
10. Given $a = 3.414$, $b = 2.875$.
11. Given $A = 84° 16'$, $a = .0033503$.
12. Given $A = 46° 23'$, $c = 5278.6$.
13. Given $a = 529.3$, $c = 902.7$.
14. Given $B = 23° 9'$, $b = 75.48$.
15. Given $A = 72° 52'$, $b = 6306$.
16. Given $B = 18° 38'$, $c = 2.5432$.
17. Given $a = .0001689$, $b = .0004761$.
18. Given $A = 31° 45'$, $a = 48.0408$.
19. Given $b = 617.57$, $c = 729.59$.
20. Given $B = 82° 6' 18''$, $a = 89.32$.
21. Given $A = 55° 43' 29''$, $c = 41518$.
22. Given $B = 31° 47' 7''$, $a = 7.23246$.
23. Given $a = 99.464$, $c = 156.819$.

24. Given $A = 43° 21' 36''$, $b = .00261751$.

25. Given $B = 79° 14' 31''$, $b = 84218.5$.

26. Given $B = 67° 39' 53''$, $c = 9537514$.

27. Given $b = 5789.72$, $c = 24916.45$.

28. Given $A = 26° 12' 24''$, $c = 469422.7$.

29. Given $B = 14° 55' 42''$, $b = .1353371$.

30. Given $a = 672.3853$, $b = 384.5038$.

Solve the following isosceles triangles, in which A and B are the equal angles, and a, b, and c the sides opposite the angles A, B, and C, respectively:

31. Given $A = 68° 57'$, $b = 350.94$.

32. Given $B = 27° 8'$, $c = 3.0892$.

33. Given $C = 84° 47'$, $b = 91032.7$.

34. Given $a = 79.2434$, $c = 106.6362$.

35. Given $A = 35° 19' 47''$, $c = .56235$.

36. Given $C = 151° 28' 52''$, $c = 9547.12$.

37. A regular pentagon is inscribed in a circle whose diameter is 35. Find the length of its side.

38. At a distance of 105 ft. from the base of a tower, the angle of elevation of its top is observed to be 38° 25′. Find its height.

39. What is the angle of elevation of the sun when a tower whose height is 103.74 ft. casts a shadow 167.38 ft. in length?

40. If the diameter of a circle is 32689, find the angle at the centre subtended by an arc whose chord is 10273.

41. If the diameter of the earth is 7912 miles, what is the distance of the remotest point of the surface visible from the summit of a mountain $1\frac{1}{4}$ miles in height?

42. Find the length of the diagonal of a regular pentagon whose side is 6.3257.

43. Find the angle of elevation of a mountain-slope which rises 238 ft. in a horizontal distance of one-eighth of a mile.

44. From the top of a lighthouse, 146 ft. above the sea, the angle of depression of a buoy is observed to be 21° 46′. Find the horizontal distance of the buoy.

45. If a pole casts a shadow which is two-thirds its own length, what is the angle of elevation of the sun?

46. A vessel is sailing due east at the rate of 7.8 miles an hour. A headland is observed to bear due north at 10.37 A.M., and 33° west of north at 12.43 P.M. Find the distance of the headland from each point of observation.

47. If a chord whose length is 41.368 subtends an arc of 145° 37′, what is the radius of the circle?

48. The length of the side of a regular octagon is 12. Find the radii of the inscribed and circumscribed circles.

49. How far from the foot of a flagpole 110 ft. high must an observer stand, so that the angle of elevation of the top of the pole may be 12°?

50. If the diagonal of a regular pentagon is 32.835, what is the radius of the circumscribed circle?

51. From the top of a tower, the angle of depression of the extremity of a horizontal base line, 1250 ft. in length measured from the foot of the tower, is observed to be 18° 36′ 29″. Find the height of the tower.

52. If the radius of a circle is 723.294, what is the length of a chord which subtends an arc of 35° 13′?

53. A regular hexagon is circumscribed about a circle whose diameter is 18. Find the length of its side.

54. From the top of a lighthouse 200 ft. above the sea, the angles of depression of two boats in line with the lighthouse are observed to be 14° and 32°, respectively. Find the distance between the boats.

55. A vessel is sailing due east at a uniform rate of speed. At 7 A.M., a lighthouse is observed bearing due north, 10.326 miles distant; and at 7.30 A.M. it bears 18° 13′ west of north. Find the rate of sailing of the vessel, and the bearing of the lighthouse at 10 A.M.

103. Care must be taken to use the Auxiliary Table for Small Angles in finding the logarithmic functions of angles between 0° and 5°, or between 85° and 90°, or the angles corresponding in the same cases.

This provides for every case which can arise in solving right triangles, except in looking out the angle corresponding to a logarithmic sine when between 85° and 90°, or a logarithmic cosine when between 0° and 5°.

We will now derive a formula for right triangles by aid of which, when b and c are given, the angle A may be determined with accuracy if it is between 85° and 90°.

By § 98, $$\cos A = \frac{b}{c}.$$

Then by (**31**), $$2\sin^2 \tfrac{1}{2} A = 1 - \cos A = 1 - \frac{b}{c} = \frac{c-b}{c}.$$

Therefore, $$\sin \tfrac{1}{2} A = \sqrt{\frac{c-b}{2c}}.$$

In like manner, $$\sin \tfrac{1}{2} B = \sqrt{\frac{c-a}{2c}}.$$

These formulæ involve the *half-angles;* hence, if the angle itself is between 85° and 90°, its half is between 42° 30′ and 45°, and the correction in seconds may in that case be found from the table with sufficient precision.

An angle between 0° and 5° may always be avoided in solving a right triangle by working with the other acute angle.

104. 1. Given $b = 1.08249$, $c = 1.08261$; find the angles.

Here A is near to 0°, and B is near to 90°, as may be determined by inspection.

We then proceed to find B by the formula of § 103.

For this purpose, we must first find a, which may be done as in § 101.

$$\begin{aligned} c + b = 2.1651;\ \log &= 0.335478 \\ c - b = .00012;\ \log &= 6.079181 - 10 \\ 2\overline{)16.414659 - 20} \\ \log a &= 8.207330 - 10 \end{aligned}$$

Whence $$a = .0161187.$$

Now to find B, we use the formula $\sin \frac{1}{2} B = \sqrt{\frac{c-a}{2c}}$.

By logarithms, $$\log \sin \tfrac{1}{2} B = \tfrac{1}{2}[\log(c-a) - \log 2c].$$

$$\begin{aligned} c - a = 1.0664913;\ \log &= 0.027957 \\ 2c = 2.16522;\ \log &= 0.335502 \\ 2\overline{)19.692455 - 20} \\ \log \sin \tfrac{1}{2} B &= 9.846228 - 10 \end{aligned}$$

Whence, $$\tfrac{1}{2} B = 44° 34' 24.7''.$$

Then, $B = 89° 8' 49.4''$, and $A = 90° - B = 0° 51' 10.6''$.

If b is small compared with c, then A is near to 90°, and should be calculated directly by aid of the formula of § 103.

EXAMPLES.

In each of the following right triangles find the angles:

2. Given $a = .0128$, $c = 152.337$.
3. Given $b = 5.81006$, $c = 5.81039$.
4. Given $c = 11527.2$, $b = 1.32$.
5. Given $a = .77$, $c = 98276.4$.
6. Given $a = 42.0098$, $c = 42.0103$.

FORMULÆ FOR THE AREA OF A RIGHT TRIANGLE.

105. CASE I. *Given the hypotenuse and an acute angle.*

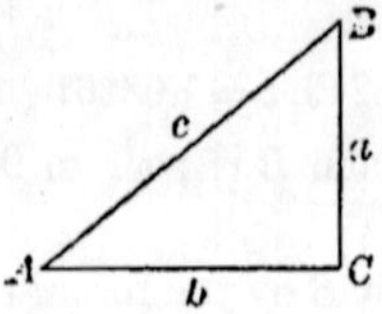

Denoting the area by K, we have by Geometry,

$$2K = ab.$$

But by § 4, $a = c \sin A$, and $b = c \cos A$.

Whence, $2K = c^2 \sin A \cos A = \frac{1}{2} c^2 \sin 2A$, by **(25)**.

Then, $$4K = c^2 \sin 2A. \quad \textbf{(38)}$$

In like manner, $$4K = c^2 \sin 2B. \quad \textbf{(39)}$$

CASE II. *Given an angle and its opposite side.*

By § 4, $b = a \cot A$.

Whence, $$2K = a \times a \cot A = a^2 \cot A. \quad \textbf{(40)}$$

In like manner, $$2K = b^2 \cot B. \quad \textbf{(41)}$$

CASE III. *Given an angle and its adjacent side.*

By § 4, $b = a \tan B$.

Whence, $$2K = a \times a \tan B = a^2 \tan B. \quad \textbf{(42)}$$

In like manner, $$2K = b^2 \tan A. \quad \textbf{(43)}$$

CASE IV. *Given the hypotenuse and another side.*

By Geometry, $b^2 = c^2 - a^2.$

Whence, $2K = ab = a\sqrt{c^2 - a^2} = a\sqrt{(c+a)(c-a)}.$ **(44)**

In like manner, $2K = b\sqrt{(c+b)(c-b)}.$ **(45)**

CASE V. *Given the two sides about the right angle.*

In this case, $2K = ab.$ **(46)**

EXAMPLES.

106. 1. Given $c = 10.3572$, $B = 74° 57' 14''$; find the area.

By (**39**), $4K = c^2 \sin 2B.$

Whence, $\log (4K) = 2 \log c + \log \sin 2B.$

$\log c = 1.015242$; multiply by $2 = 2.030484$

$2B = 149° 54' 28''$; $\log \sin = 9.700178 - 10$

$\log (4K) = 1.730662$

$4K = 53.7851$

Dividing by 4, $K = 13.4463.$

Note. To find $\log \sin 149° 54' 28''$, take either $\log \cos 59° 54' 28''$, or $\log \sin 30° 5' 32''$. (See Introduction to Tables, page viii.)

Find the areas of the following right angles:

2. Given $A = 19° 36'$, $a = 2.2178$.
3. Given $B = 24° 7' 48''$, $a = .8213$.
4. Given $a = 149.417$, $b = 76.292$.
5. Given $b = .305694$, $c = .660156$.
6. Given $A = 30° 56' 19''$, $c = 192.035$.
7. Given $A = 78° 42' 53''$, $b = .0520281$.
8. Given $a = .932368$, $c = 4.786723$.
9. Given $B = 72° 18' 27''$, $c = 27.28338$.
10. Given $B = 49° 25' 34''$, $b = .3375494$.

VII. GENERAL PROPERTIES OF TRIANGLES.

107. *In any triangle, the sides are proportional to the sines of their opposite angles.*

I. To prove $a : b = \sin A : \sin B.$ **(47)**

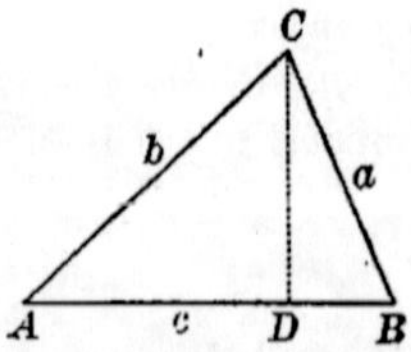

Fig. 1.

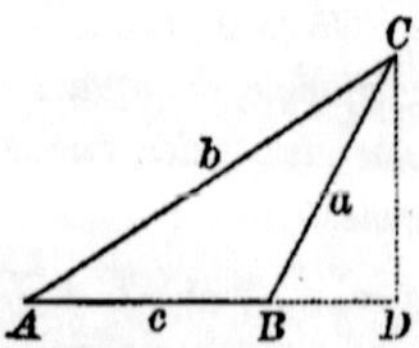

Fig. 2.

There will be two cases, according as the angles A and B are both acute (Fig. 1), or one of them obtuse (Fig. 2).

In each case, draw CD perpendicular to AB.

Then in each figure, $CD = b \sin A$ (§ 4).

Also in Fig. 1, $CD = a \sin B.$

And in Fig. 2, $CD = a \sin CBD$

$$= a \sin (180° - B) = a \sin B \text{ (§ 33)}.$$

Then in either case, $b \sin A = a \sin B.$

Whence by the theory of proportion,

$$a : b = \sin A : \sin B.$$

In like manner, $b : c = \sin B : \sin C,$ **(48)**

and $c : a = \sin C : \sin A.$ **(49)**

108. *In any triangle, the sum of any two sides is to their difference as the tangent of half the sum of the opposite angles is to the tangent of half their difference.*

By (**47**), $a : b = \sin A : \sin B.$

Whence by composition and division,

$$a + b : a - b = \sin A + \sin B : \sin A - \sin B.$$

Or, $$\frac{a+b}{a-b} = \frac{\sin A + \sin B}{\sin A - \sin B}.$$

But, $$\frac{\sin A + \sin B}{\sin A - \sin B} = \frac{\tan \frac{1}{2}(A+B)}{\tan \frac{1}{2}(A-B)},$$ by (**21**).

Whence, $$\frac{a+b}{a-b}=\frac{\tan\frac{1}{2}(A+B)}{\tan\frac{1}{2}(A-B)}. \qquad (50)$$

In like manner, $$\frac{b+c}{b-c}=\frac{\tan\frac{1}{2}(B+C)}{\tan\frac{1}{2}(B-C)}, \qquad (51)$$

and $$\frac{c+a}{c-a}=\frac{\tan\frac{1}{2}(C+A)}{\tan\frac{1}{2}(C-A)}. \qquad (52)$$

109. *In any triangle, the square of any side is equal to the sum of the squares of the other two sides, minus twice their product into the cosine of their included angle.*

I. To prove $$a^2=b^2+c^2-2\,bc\cos A. \qquad (53)$$

CASE I. *When the included angle A is acute.* (Figures of § 107.)

There will be two cases, according as the angle B is acute (Fig. 1), or obtuse (Fig. 2).

Then in Fig. 1, $BD=c-AD$, and in Fig. 2, $BD=AD-c$.

Squaring, we have in either case,

$$\overline{BD}^2=\overline{AD}^2+c^2-2\,c\times AD.$$

Adding $\overline{CD}^2$ to both members,

$$\overline{BD}^2+\overline{CD}^2=\overline{AD}^2+\overline{CD}^2+c^2-2\,c\times AD.$$

But $\overline{BD}^2+\overline{CD}^2=a^2$, and $\overline{AD}^2+\overline{CD}^2=b^2$.

Also, by § 4, $AD=b\cos A$.

Whence, $a^2=b^2+c^2-2\,bc\cos A$.

CASE II. *When the included angle A is obtuse.*

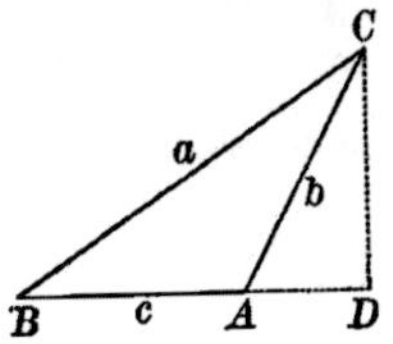

Draw CD perpendicular to AB.

We have $BD=AD+c$.

Squaring, and adding $\overline{CD}^2$ to both members,

$$\overline{BD}^2+\overline{CD}^2=\overline{AD}^2+\overline{CD}^2+c^2+2\,c\times AD.$$

But $\overline{BD}^2 + \overline{CD}^2 = a^2$, and $\overline{AD}^2 + \overline{CD}^2 = b^2$.

And by § 4, $AD = b \cos CAD = b \cos (180° - A) = -b \cos A$ (§ 33).

Whence, $a^2 = b^2 + c^2 - 2\,bc \cos A$.

In like manner, $b^2 = c^2 + a^2 - 2\,ca \cos B$, **(54)**

and $c^2 = a^2 + b^2 - 2\,ab \cos C$. **(55)**

110. *To express the cosines of the angles of a triangle in terms of the sides of the triangle.*

By **(53)**, $a^2 = b^2 + c^2 - 2\,bc \cos A$.

Transposing, $2\,bc \cos A = b^2 + c^2 - a^2$.

Whence, $\cos A = \dfrac{b^2 + c^2 - a^2}{2\,bc}$. **(56)**

In like manner, $\cos B = \dfrac{c^2 + a^2 - b^2}{2\,ca}$, **(57)**

and $\cos C = \dfrac{a^2 + b^2 - c^2}{2\,ab}$. **(58)**

111. *To express the sines, cosines, and tangents of the half-angles of a triangle in terms of the sides of the triangle.*

By **(56)**, $1 - \cos A = 1 - \dfrac{b^2 + c^2 - a^2}{2\,bc} = \dfrac{a^2 - b^2 + 2\,bc - c^2}{2\,bc}$.

Whence, by **(31)**, $2 \sin^2 \frac{1}{2} A = \dfrac{a^2 - (b - c)^2}{2\,bc}$.

Or, $\sin^2 \frac{1}{2} A = \dfrac{(a - b + c)(a + b - c)}{4\,bc}$.

Denoting the sum of the sides, $a + b + c$, by $2\,s$, we have

$$a - b + c = (a + b + c) - 2\,b = 2\,s - 2\,b = 2(s - b),$$

and $$a + b - c = (a + b + c) - 2\,c = 2\,s - 2\,c = 2(s - c).$$

Whence, $\sin^2 \frac{1}{2} A = \dfrac{4(s - b)(s - c)}{4\,bc}$.

Or, $\sin \frac{1}{2} A = \sqrt{\dfrac{(s - b)(s - c)}{bc}}$. **(59)**

In like manner, $$\sin \tfrac{1}{2} B = \sqrt{\frac{(s-c)(s-a)}{ca}}, \quad (60)$$

and $$\sin \tfrac{1}{2} C = \sqrt{\frac{(s-a)(s-b)}{ab}}. \quad (61)$$

Again, by (56), $$1 + \cos A = 1 + \frac{b^2 + c^2 - a^2}{2bc} = \frac{b^2 + 2bc + c^2 - a^2}{2bc}.$$

Whence, by (32), $$2\cos^2 \tfrac{1}{2} A = \frac{(b+c)^2 - a^2}{2bc}.$$

Or, $$\cos^2 \tfrac{1}{2} A = \frac{(b+c+a)(b+c-a)}{4bc}.$$

But, $b + c + a = 2s$;

and $b + c - a = (b + c + a) - 2a = 2(s - a)$.

Whence, $$\cos^2 \tfrac{1}{2} A = \frac{4s(s-a)}{4bc}.$$

Or, $$\cos \tfrac{1}{2} A = \sqrt{\frac{s(s-a)}{bc}}. \quad (62)$$

In like manner, $$\cos \tfrac{1}{2} B = \sqrt{\frac{s(s-b)}{ca}}, \quad (63)$$

and $$\cos \tfrac{1}{2} C = \sqrt{\frac{s(s-c)}{ab}}. \quad (64)$$

Dividing (59) by (62), we have, by (4),

$$\tan \tfrac{1}{2} A = \sqrt{\frac{(s-b)(s-c)}{bc}} \sqrt{\frac{bc}{s(s-a)}} = \sqrt{\frac{(s-b)(s-c)}{s(s-a)}}. \quad (65)$$

In like manner, $$\tan \tfrac{1}{2} B = \sqrt{\frac{(s-c)(s-a)}{s(s-b)}}, \quad (66)$$

and $$\tan \tfrac{1}{2} C = \sqrt{\frac{(s-a)(s-b)}{s(s-c)}}. \quad (67)$$

Note. Since each angle of a triangle is less than 180°, its half is less than 90°; hence, the *positive sign* must be taken before the radical in each formula of § 111.

FORMULÆ FOR THE AREA OF AN OBLIQUE TRIANGLE.

112. Case I. *Given two sides and their included angle.*

I. When the given parts are b, c, and A.

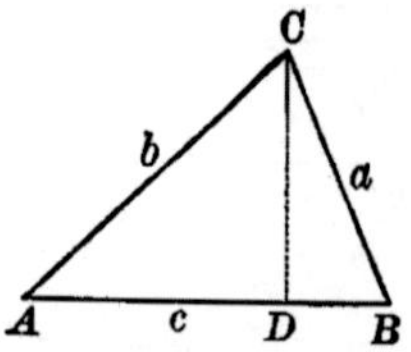

FIG. 1.

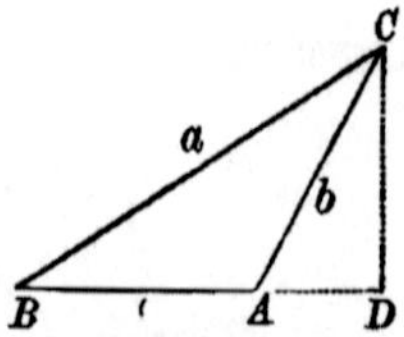

FIG. 2.

There will be two cases, according as A is acute (Fig. 1), or obtuse (Fig. 2).

In each case, draw CD perpendicular to AB.

Then denoting the area by K, we have by Geometry,

$$2K = c \times CD.$$

But in Fig. 1, $CD = b \sin A$ (§ 4).

And in Fig. 2, $CD = b \sin CAD$

$$= b \sin (180° - A) = b \sin A \text{ (§ 33)}.$$

Then in either case, $2K = bc \sin A$. **(68)**

In like manner, $2K = ca \sin B$, **(69)**

and $2K = ab \sin C$. **(70)**

CASE II. *Given one side and all the angles.*

I. When the given parts are a, A, B, and C.

By **(70)**, $2K = ab \sin C$.

But by **(47)**, $\frac{b}{a} = \frac{\sin B}{\sin A}$, or $b = \frac{a \sin B}{\sin A}$.

Whence, $2K = a \times \frac{a \sin B}{\sin A} \times \sin C = \frac{a^2 \sin B \sin C}{\sin A}$. **(71)**

In like manner, $2K = \frac{b^2 \sin C \sin A}{\sin B}$, **(72)**

and $2K = \frac{c^2 \sin A \sin B}{\sin C}$. **(73)**

CASE III. *Given the three sides.*

By **(68)**, $2K = bc \sin A = 2bc \sin \frac{1}{2} A \cos \frac{1}{2} A$, by **(25)**.

Dividing by 2, and substituting the values of $\sin \frac{1}{2} A$ and $\cos \frac{1}{2} A$ from **(59)** and **(62)**, we have,

$$K = bc\sqrt{\frac{(s-b)(s-c)}{bc}}\sqrt{\frac{s(s-a)}{bc}} = \sqrt{s(s-a)(s-b)(s-c)}. \quad \textbf{(74)}$$

VIII. SOLUTION OF OBLIQUE TRIANGLES.

113. In the solution of oblique triangles, we may distinguish four cases.

114. CASE I. *Given a side and any two angles.*

The third angle may be found by Geometry, and then by aid of § 107 the remaining sides may be calculated.

The triangle is always possible for any values of the given elements, provided the sum of the given angles is less than 180°.

1. Given $b = 20.24$, $A = 103° 36'$, $B = 19° 21'$; find C, a, and c.

We have, $C = 180° - (A + B) = 180° - 122° 57' = 57° 3'$.

By (**47**), $$\frac{a}{b} = \frac{\sin A}{\sin B}, \text{ and } \frac{c}{b} = \frac{\sin C}{\sin B}.$$

Then, $$a = b \sin A \csc B, \text{ and } c = b \sin C \csc B.$$

Whence, $$\log a = \log b + \log \sin A + \log \csc B,$$

and $$\log c = \log b + \log \sin C + \log \csc B.$$

$$\begin{aligned} \log b &= 1.306211 \\ \log \sin A &= 9.987649 - 10 \\ \log \csc B &= 0.479729 \\ \hline \log a &= 1.773589 \\ a &= 59.3730. \end{aligned} \qquad \begin{aligned} \log b &= 1.306211 \\ \log \sin C &= 9.923837 - 10 \\ \log \csc B &= 0.479729 \\ \hline \log c &= 1.709777 \\ c &= 51.2598. \end{aligned}$$

Note. To find log sin 103° 36′, take either log cos 13° 36′, or log sin 76° 24′. To find the log cosecant of an angle, subtract the log sine from 10 – 10. (See Introduction to Tables, page viii.)

EXAMPLES.

Solve the following triangles:

2. Given $a = 180$, $A = 38°$, $B = 75° 43'$.
3. Given $b = .82$, $B = 51° 42' 37''$, $C = 109° 17' 23''$.
4. Given $c = 24.637$, $A = 83° 39'$, $B = 38° 56'$.
5. Given $b = .06708$, $A = 26° 10' 45''$, $C = 44° 35' 12''$.
6. Given $a = 5.0454$, $B = 98° 8' 26''$, $C = 21° 51' 34''$.
7. Given $c = 4592.36$, $A = 74° 27'$, $C = 61°$.
8. Given $c = .93109$, $A = 15° 34' 9''$, $C = 123° 29' 46''$.
9. Given $b = 3.67683$, $A = 67° 21' 54''$, $B = 57° 48' 8''$.
10. Given $a = 71396.72$, $B = 42° 55' 13''$, $C = 16° 4' 57''$.

115. CASE II. *Given two sides and their included angle.*

Since one angle is known, the sum of the other two angles may be found; and then their difference may be calculated by aid of § 108.

Knowing the sum and difference of the angles, the angles themselves may be found; and then the remaining side may be computed as in Case I.

The triangle is possible for any values of the data.

1. Given $a = 82$, $c = 167$, $B = 98° 14'$; find A, C, and b.

By Geometry, $C + A = 180° - B = 81° 46'$.

By (52), $$\frac{c+a}{c-a} = \frac{\tan\frac{1}{2}(C+A)}{\tan\frac{1}{2}(C-A)}.$$

Or, $$\tan\tfrac{1}{2}(C-A) = \frac{c-a}{c+a}\tan\tfrac{1}{2}(C+A).$$

Then, $\log\tan\frac{1}{2}(C-A) = \log(c-a) + \text{colog}(c+a) + \log\tan\frac{1}{2}(C+A)$.

$$\begin{array}{lr} c - a = 85 & \log = 1.929419 \\ c + a = 249 & \text{colog} = 7.603801 - 10 \\ \frac{1}{2}(C+A) = 40° 53' & \log\tan = 9.937377 - 10 \\ \hline & \log\tan\frac{1}{2}(C-A) = 9.470597 - 10 \\ & \frac{1}{2}(C-A) = 16° 27' 49.8''. \end{array}$$

Therefore, $C = \frac{1}{2}(C+A) + \frac{1}{2}(C-A) = 57° 20' 49.8''$,

and $A = \frac{1}{2}(C+A) - \frac{1}{2}(C-A) = 24° 25' 10.2''$.

To find the remaining side, we have by (47),

$$b = \frac{a \sin B}{\sin A} = a \sin B \csc A.$$

Whence, $\log b = \log a + \log\sin B + \log\csc A$.

$$\begin{array}{r} \log a = 1.913814 \\ \log\sin B = 9.995501 - 10 \\ \log\csc A = 0.383615 \\ \hline \log b = 2.292930 \\ b = 196.305. \end{array}$$

EXAMPLES.

Solve the following triangles:

2. Given $a = 67$, $c = 33$, $B = 36°$.
3. Given $a = 886$, $b = 747$, $C = 71° 54'$.
4. Given $b = 4.102$, $c = 4.549$, $A = 62° 9' 38$.

5. Given $a = .5953$, $b = .9639$, $C = 134°$.

6. Given $b = 1292.1$, $c = 286.3$, $A = 27° 13'$.

7. Given $a = 7.48$, $c = 12.409$, $B = 83° 26' 52''$.

8. Given $a = 93.273$, $b = 81.512$, $C = 58°$.

9. Given $b = .0261579$, $c = .0608657$, $A = 115° 42'$.

10. Given $a = 35384.82$, $c = 57946.34$, $B = 19° 37' 25''$.

116. Case III. *Given the three sides.*

The angles might be calculated by the formulæ of § 110; but as these are not adapted to logarithmic computation, it is usually more convenient to use the formulæ of § 111.

Each of the three angles should be computed trigonometrically, for we then have a check on the work, since their sum should be 180°.

If all the angles are to be computed, the *tangent* formulæ are the most convenient, since only four different logarithms are required. If but one angle is required, the *cosine* formula will be found to involve the least work.

The triangle is possible for any values of the data, provided no side is greater than the sum of the other two.

If all the angles are required, and the tangent formulæ are used, it is convenient to modify them as follows; by (**65**),

$$\tan \tfrac{1}{2} A = \sqrt{\frac{(s-a)(s-b)(s-c)}{s(s-a)^2}} = \frac{1}{s-a}\sqrt{\frac{(s-a)(s-b)(s-c)}{s}}.$$

Denoting $\sqrt{\dfrac{(s-a)(s-b)(s-c)}{s}}$ by r, we have

$$\tan \tfrac{1}{2} A = \frac{r}{s-a}.$$

In like manner, $\tan \frac{1}{2} B = \dfrac{r}{s-b}$, and $\tan \frac{1}{2} C = \dfrac{r}{s-c}$.

1. Given $a = 2.51$, $b = 2.79$, $c = 2.33$; find A, B, and C.

Here, $$2s = a + b + c = 7.63.$$

Whence, $s = 3.815$, $s - a = 1.305$, $s - b = 1.025$, $s - c = 1.485$.

We have, $\log r = \frac{1}{2}[\log(s-a) + \log(s-b) + \log(s-c) + \text{colog } s]$.

Also,
$$\log \tan \tfrac{1}{2} A = \log r - \log(s-a),$$
$$\log \tan \tfrac{1}{2} B = \log r - \log(s-b),$$
and
$$\log \tan \tfrac{1}{2} C = \log r - \log(s-c).$$

$$
\begin{aligned}
\log(s-a) &= 0.115611 \\
\log(s-b) &= 0.010724 \\
\log(s-c) &= 0.171726 \\
\text{colog}\, s &= 9.418505-10 \\
2)\overline{19.716566-20} \\
\log r &= 9.858283-10 \\
\log(s-a) &= 0.115611 \\
\log\tan\tfrac{1}{2}A &= 9.742672-10 \\
\tfrac{1}{2}A &= 28^\circ\,56'\,22.7''. \\
A &= 57^\circ\,52'\,45.4''.
\end{aligned}
$$

$$
\begin{aligned}
\log r &= 9.858283-10 \\
\log(s-b) &= 0.010724 \\
\log\tan\tfrac{1}{2}B &= 9.847559-10 \\
\tfrac{1}{2}B &= 35^\circ\,8'\,40.9''. \\
B &= 70^\circ\,17'\,21.8''.
\end{aligned}
$$

$$
\begin{aligned}
\log r &= 9.858283-10 \\
\log(s-c) &= 0.171726 \\
\log\tan\tfrac{1}{2}C &= 9.686557-10 \\
\tfrac{1}{2}C &= 25^\circ\,54'\,56.2''. \\
C &= 51^\circ\,49'\,52.4''.
\end{aligned}
$$

Check, $A+B+C=179^\circ\,59'\,59.6''$.

2. Given $a=7$, $b=11$, $c=9.6$; find B.

By (63), $$\cos\tfrac{1}{2}B=\sqrt{\frac{s(s-b)}{ca}}.$$

Whence, $\log\cos\frac{1}{2}B=\frac{1}{2}[\log s+\log(s-b)+\text{colog}\,c+\text{colog}\,a]$.

Here, $2s=a+b+c=27.6$; whence, $s=13.8$, and $s-b=2.8$.

$$
\begin{aligned}
\log s &= 1.139879 \\
\log(s-b) &= 0.447158 \\
\text{colog}\, c &= 9.017729-10 \\
\text{colog}\, a &= 9.154902-10 \\
2)\overline{19.759668-20} \\
\log\cos\tfrac{1}{2}B &= 9.879834-10
\end{aligned}
$$

$\frac{1}{2}B=40^\circ\,41'\,11.5''$, and $B=81^\circ\,22'\,23.0''$.

EXAMPLES.

Solve the following triangles:

3. Given $a=2$, $b=3$, $c=4$.

4. Given $a=5$, $b=7$, $c=6$.

5. Given $a=10$, $b=9$, $c=8$.

6. Given $a=5.6$, $b=4.3$, $c=4.9$.

7. Given $a=.85$, $b=.92$, $c=.78$.

8. Given $a=61.3$, $b=84.7$, $c=47.6$.

9. Given $a=705$, $b=562$, $c=639$; find A.

10. Given $a=.0291$, $b=.0184$, $c=.0358$; find B.

11. Given $a=3019$, $b=6731$, $c=4228$; find C.

117. Case IV. *Given two sides and the angle opposite to one of them.*

It was stated in § 97 that a triangle is, in general, completely determined when three of its elements are known, provided one of them is a side. The only exceptions occur in Case IV.

To illustrate, let us consider the following example:

Given $a = 52.1$, $b = 61.2$, $A = 31° 26'$; find B, C, and c.

By (47), $$\frac{\sin B}{\sin A} = \frac{b}{a}, \text{ or } \sin B = \frac{b \sin A}{a}.$$

Whence, $$\log \sin B = \log b + \text{colog}\, a + \log \sin A.$$

$$\begin{aligned} \log b &= 1.786751 \\ \text{colog}\, a &= 8.283162 - 10 \\ \log \sin A &= 9.717259 - 10 \\ \hline \log \sin B &= 9.787172 - 10 \\ B &= 37° 46' 37.9'', \text{ from the table.} \end{aligned}$$

But in determining the angle corresponding, attention must be paid to the fact that an angle and its supplement have the same sine (§ 33).

Hence, another value of B will be $180° - 37° 46' 37.9''$, or $142° 13' 22.1''$; and calling these values B_1 and B_2, we have

$$B_1 = 37° 46' 37.9'', \text{ and } B_2 = 142° 13' 22.1''.$$

Note. The reason for this ambiguity is at once apparent when we attempt to construct the triangle from the data.

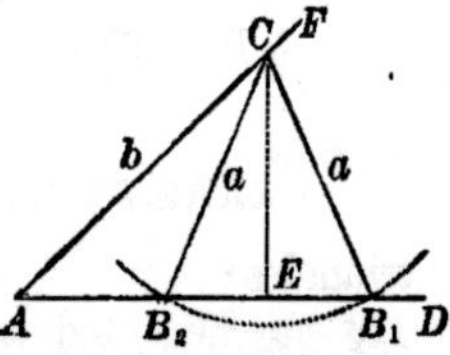

We first lay off the angle DAF equal to $31° 26'$, and on AF take $AC = 61.2$. With C as a centre, and a radius equal to 52.1, describe an arc cutting AD at B_1 and B_2. Then either of the triangles AB_1C or AB_2C satisfies the given conditions.

The two values of B which were obtained are the values of the angles AB_1C and AB_2C, respectively; and it is evident geometrically that these angles are supplementary.

To complete the solution, denote the angles ACB_1 and ACB_2 by C_1 and C_2, and the sides AB_1 and AB_2 by c_1 and c_2.

Then, $C_1 = 180° - (A + B_1) = 180° - 69° 12' 37.9'' = 110° 47' 22.1''$,

and $C_2 = 180° - (A + B_2) = 180° - 173° 39' 22.1'' = 6° 20' 37.9''$.

Again, by (**49**), $\frac{c_1}{a}=\frac{\sin C_1}{\sin A}$, and $\frac{c_2}{a}=\frac{\sin C_2}{\sin A}$.

Whence, $c_1 = a \sin C_1 \csc A$, and $c_2 = a \sin C_2 \csc A$.

$\log a = 1.716838$	$\log a = 1.716838$
$\log \sin C_1 = 9.970761 - 10$	$\log \sin C_2 = 9.043343 - 10$
$\log \csc A = 0.282741$	$\log \csc A = 0.282741$
$\log c_1 = 1.970340$	$\log c_2 = 1.042922$
$c_1 = 93.3985.$	$c_2 = 11.0388.$

118. Whenever an angle of an oblique triangle is determined from its *sine*, both the acute and obtuse values must be retained as solutions, unless one of them can be shown by other considerations to be inadmissible; and hence there may sometimes be two solutions, sometimes one, and sometimes none, in an example under Case IV.

1. Let the data be a, b, and A, and suppose $b < a$.

By Geometry, B must be $< A$; hence, only the *acute* value of B can be taken; in this case there is but *one* solution.

2. Let the data be a, b, and A, and suppose $b > a$.

Since B must be $> A$, the triangle is impossible unless A is acute.

Again, since $\frac{\sin B}{\sin A}=\frac{b}{a}$, and b is $> a$, $\sin B$ is $> \sin A$.

Hence, both the acute and obtuse values of B are $> A$, and there are *two* solutions, except in the following cases:

If $\log \sin B = 0$, then $\sin B = 1$ (§ 72), and $B = 90°$, and the triangle is a *right* triangle; if $\log \sin B$ is *positive*, then $\sin B$ is >1, and the triangle is impossible.

119. The results of § 118 may be stated as follows:

If, of the given sides, that adjacent to the given angle is the *less*, there is but *one* solution, corresponding to the *acute* value of the opposite angle.

If the side adjacent to the given angle is the *greater*, there are *two* solutions, unless the log sine of the opposite angle is 0 or positive; in which cases there are *one* solution (a *right* triangle), and *no* solution, respectively.

120. We will illustrate the above points by examples:

1. Given $a = 7.42$, $b = 3.39$, $A = 105°13'$; find B.

Since b is $< a$, there is but *one* solution, corresponding to the *acute* value of B.

By (47), $$\sin B = \frac{b \sin A}{a}.$$

Whence, $$\log \sin B = \log b + \operatorname{colog} a + \log \sin A.$$

$$\begin{aligned} \log b &= 0.530200 \\ \operatorname{colog} a &= 9.129596 - 10 \\ \log \sin A &= 9.984500 - 10 \\ \hline \log \sin B &= 9.644296 - 10 \\ B &= 26^\circ\, 9'\, 30.5''. \end{aligned}$$

2. Given $b = 3$, $c = 2$, $C = 100^\circ$; find B.

Since b is $> c$, and C is obtuse, the triangle is impossible.

3. Given $a = 22.7643$, $c = 50$, $A = 27^\circ\, 5'$; find C.

We have, $$\sin C = \frac{c \sin A}{a}.$$

$$\begin{aligned} \log c &= 1.698970 \\ \operatorname{colog} a &= 8.642746 - 10 \\ \log \sin A &= 9.658284 - 10 \\ \hline \log \sin C &= 0.000000 \end{aligned}$$

Therefore, $\sin C = 1$, and $C = 90^\circ$.

Here there is but one solution; a *right* triangle.

4. Given $a = .83$, $b = .715$, $B = 61^\circ\, 47'$; find A.

We have, $$\sin A = \frac{a \sin B}{b}.$$

$$\begin{aligned} \log a &= 9.919078 - 10 \\ \operatorname{colog} b &= 0.145694 \\ \log \sin B &= 9.945058 - 10 \\ \hline \log \sin A &= 0.009830 \end{aligned}$$

Since $\log \sin A$ is positive, the triangle is impossible.

EXAMPLES.

121. Solve the following triangles:

1. Given $a = 5.98$, $b = 3.59$, $A = 63^\circ\, 50'$.
2. Given $b = 74.1$, $c = 64.2$, $C = 27^\circ\, 18'$.
3. Given $b = .2337$, $c = .0982$, $B = 108^\circ$.
4. Given $a = 4.254$, $c = 4.536$, $C = 37^\circ\, 9'$.
5. Given $a = .2789$, $b = .2271$, $B = 65^\circ\, 38'$.

6. Given $a = 60.935$, $c = 76.097$, $A = 133° 41'$.

7. Given $b = 74.8067$, $c = 98.7385$, $C = 81° 47'$.

8. Given $a = 9.51987$, $c = 11$, $A = 59° 56'$.

9. Given $b = 4.521$, $c = 5.03$, $B = 40° 32' 7''$.

10. Given $a = 186.82$, $b = 394.2$, $B = 114° 29' 51''$.

11. Given $b = 5143.4$, $c = 4795.56$, $C = 72° 53' 38''$.

12. Given $a = .860619$, $c = .635761$, $A = 19° 12' 43''$.

13. Given $a = 139.27$, $b = 195.9716$, $A = 45° 17' 20''$.

14. Given $a = .32163$, $c = .27083$, $C = 52° 24' 16''$.

15. Given $b = 91139.04$, $c = 80640.37$, $B = 126° 5' 34''$.

AREA OF AN OBLIQUE TRIANGLE.

122. 1. Given $a = 18.063$, $A = 96° 30' 15''$, $B = 35° 0' 13''$; find K.

By (71), $$2K = \frac{a^2 \sin B \sin C}{\sin A} = a^2 \sin B \sin C \csc A.$$

Whence, $\log(2K) = 2\log a + \log\sin B + \log\sin C + \log\csc A$.

From the data, $C = 180° - (A + B) = 48° 29' 32''$.

$$\begin{aligned}
\log a = 1.256790;\ \text{multiply by } 2 &= 2.513580 \\
\log\sin B &= 9.758630 - 10 \\
\log\sin C &= 9.874404 - 10 \\
\log\csc A &= 0.002804 \\
\hline
\log(2K) &= 2.149418 \\
2K &= 141.065. \\
K &= 70.533.
\end{aligned}$$

EXAMPLES.

Find the areas of the following triangles:

2. Given $a = 38.09$, $c = 11.2$, $B = 67° 55'$.

3. Given $a = 5$, $b = 8$, $c = 6$.

4. Given $b = 6.074$, $A = 70° 39'$, $B = 56° 23'$.

5. Given $b = 761.86$, $c = 526.02$, $A = 124° 6' 13''$.

6. Given $a = 97$, $b = 83$, $c = 71$.

7. Given $a = 1.9375$, $A = 43° 18'$, $B = 29° 47' 36''$.

8. Given $b = .439592$, $A = 62° 40' 8''$, $C = 54° 32' 25''$.

9. Given $a = 39.5$, $b = 44.8$, $c = 52.3$.

10. Given $a = .804639$, $c = .357173$, $B = 18° 11' 49''$.

11. Given $c = 95.86157$, $B = 115° 24' 52''$, $C = 32° 57' 21''$.

12. Given $a = .02409481$, $b = .02763834$, $C = 81° 9' 34''$.

13. Given $a = 7.825$, $b = 6.592$, $c = 9.643$.

MISCELLANEOUS EXAMPLES.

123. 1. From a point in the same horizontal plane with the base of a tower, the angle of elevation of its top is 52° 39′, and from a point 100 ft. further away it is 35° 16′. Find the height of the tower, and its distance from each point of observation.

2. One side of a parallelogram is 56, and the angles between this side and the diagonals are 31° 14′ and 45° 37′. Find all the sides of the parallelogram.

3. In a field *ABCD*, the sides *AB*, *BC*, *CD*, and *DA* are 155, 236, 252, and 105 rods, respectively, and the diagonal *AC* is 311 rods. Find the area of the field.

4. The area of a triangle is 1356, and two of its sides are 53 and 69. Find the angle between them.

5. From the top of a bluff, the angles of depression of two posts in the plain below, in line with the observer and 1000 ft. apart, are found to be 27° 40′ and 9° 33′, respectively. Find the height of the bluff above the plain.

6. The parallel sides of a trapezoid are 86 and 138, and the angles at the extremities of the latter are 53° 49′ and 67° 55′. Find the non-parallel sides.

7. Two trains start at the same time from the same point, and move along straight railways, which intersect at an angle of 74° 30′, at the rates of 30 and 45 miles an hour, respectively. How far apart are they at the end of 45 minutes?

8. Two sides of a triangle are .5623 and .4977, and the difference of the angles opposite these sides is 15° 48′ 32″. Solve the triangle.

9. Two yachts start at the same time from the same point, and sail, one due north at the rate of 10.44 miles an hour, and the other due north-east at the rate of 7.71 miles an hour. What is the bearing of the first yacht from the second at the end of half an hour?

10. A vessel is sailing due south-west at the rate of 8 miles an hour. At 10.30 A.M., a lighthouse is observed to bear 30° west of north, and at 12.15 P.M., it is observed to bear 15° east of north. Find the distance of the lighthouse from each position of the vessel.

11. Two sides of a parallelogram are 65 and 133, and one of the diagonals is 159. Find the angles of the parallelogram, and the other diagonal.

12. To find the distance of an inaccessible object A from a position B, I measure a line BC, 208.3 ft. in length. The angles ABC and ACB are measured, and found to be 126° 35′ and 31° 48′, respectively. Find the distance AB.

13. The diagonals of a parallelogram are 81 and 106, and the angle between them is 29° 18′. Find the sides and angles of the parallelogram.

14. A flagpole 40 ft. high stands on the top of a tower. From a position near the base of the tower, the angles of elevation of the top and bottom of the pole are 38° 53′ and 20° 18′, respectively. Find the distance and height of the tower.

15. AD and BC are the parallel sides of a trapezoid $ABCD$; the sides AB and BC are 7.8 and 9.4, respectively, and the angles B and C are 113° 47′ and 125° 34′, respectively. Find AD and CD.

16. A surveyor observes that his position A is exactly in line with two inaccessible objects B and C. He measures a line AD 500 ft. long, making the angle $BAD = 60°$, and at D observes the angles ADB and BDC to be 40° and 60°, respectively. Find the distance BC.

17. A side of a parallelogram is 48, a diagonal is 73, and the angle between the diagonals, opposite the given side, is 98° 6′. Find the other diagonal and the other side.

18. To find the distance between two buoys A and B, I measure a base-line CD on the shore, 150 ft. long. At the point C the angles ACD and BCD are measured, and found to be 95° and 70°, respectively; and at D the angles BDC and ADC are found to be 83° and 30°, respectively. Find the distance between the buoys.

19. The sides AB, BC, and CD, of a quadrilateral $ABCD$ are 38, 55, and 42, respectively, and the angles B and C are 132° 56′ and 98° 29′, respectively. Find the side AD, and the angles A and D.

20. The sides AB, BC, and DA of a field $ABCD$ are 37, 63, and 20 rods, respectively, and the diagonals AC and BD are 75 and 42 rods, respectively. Find the area of the field.

IX. CUBIC EQUATIONS.

124. We know, by Algebra, that a cubic equation can always be transformed into another in which the term containing the square of the unknown quantity shall be wanting.

Thus, if the equation is $x^3+px^2+qx+r=0$, putting $x=y-\frac{p}{3}$, we have

$$y^3-py^2+\frac{p^2y}{3}-\frac{p^3}{27}+py^2-\frac{2p^2y}{3}+\frac{p^3}{9}+qy-\frac{pq}{3}+r=0;$$

or,
$$y^3+y\left(q-\frac{p^2}{3}\right)+\frac{2p^3}{27}-\frac{pq}{3}+r=0;$$

which is in the required form.

125. Cardan's Method enables us to solve any cubic equation of the form $x^3+ax+b=0$, except in the case where a is negative, and $\frac{a^3}{27}$ numerically $>\frac{b^2}{4}$.

In this case, it is possible to find the roots by Trigonometry.

126. Trigonometric Solution of Cubic Equations.

To solve the equation

$$x^3-ax-b=0,$$

where a is positive, and $\frac{a^3}{27}>\frac{b^2}{4}$.

Putting $x=2m\cos A$, the equation becomes

$$8m^3\cos^3 A-2am\cos A-b=0, \text{ or } 4\cos^3 A-\frac{a}{m^2}\cos A-\frac{b}{2m^3}=0.$$

But by (**36**), $\quad 4\cos^3 A=\cos 3A+3\cos A.$

Whence, $\quad \cos 3A+3\cos A-\frac{a}{m^2}\cos A-\frac{b}{2m^3}=0.$

Or, $$\cos 3A+\left(3-\frac{a}{m^2}\right)\cos A=\frac{b}{2m^3}. \qquad \text{(A)}$$

We may take m so that $3-\frac{a}{m^2}=0$; then, $3m^2=a$, and $m=\sqrt{\frac{a}{3}}$. (B)

Then (A) becomes $\quad \cos 3A=\frac{b}{2m^3}.$

Substituting in this the value of m from (B), we have

$$\cos 3A = \frac{b}{2}\sqrt{\frac{27}{a^3}}. \qquad (C)$$

Since, by hypothesis, $\frac{b^2}{4} < \frac{a^3}{27}$, we have $\frac{b^2}{4} \times \frac{27}{a^3} < 1$.

Taking the square root of each member of the inequality, $\frac{b}{2}\sqrt{\frac{27}{a^3}} < 1$.

Hence, the value of $3A$ in (C) is possible, since its cosine is < 1.

Let z be the least positive angle whose cosine is equal to $\frac{b}{2}\sqrt{\frac{27}{a^3}}$.

Then, one value of $3A$ is z; and all its values are given by the expression $2n\pi \pm z$ (§ 62), where n is 0 or any positive or negative integer.

Whence, $\qquad \cos A = \cos \frac{1}{3}(2n\pi \pm z)$.

Now let $n = 3q + n'$, where q is 0 or any positive or negative integer, and $n' = 0$ or ± 1; then,

$$\cos A = \cos\frac{(6q + 2n')\pi \pm z}{3} = \cos\left[2q\pi + \frac{2n'\pi \pm z}{3}\right] = \cos\frac{2n'\pi \pm z}{3};$$

for by § 21, any multiple of 360° may be added to, or subtracted from, an angle, without altering its functions.

Putting $n' = 0$, 1, and -1, we have

$$\cos A = \cos\left(\pm\frac{z}{3}\right),\ \cos\frac{2\pi \pm z}{3},\ \text{or}\ \cos\frac{-2\pi \pm z}{3} = \cos\frac{z}{3}\ \text{or}\ \cos\frac{2\pi \pm z}{3};$$

for by § 29, the cosine of the angle $(-A)$ is equal to the cosine of A.

But $x = 2m\cos A$; and hence the three values of x are

$$2\sqrt{\frac{a}{3}}\cos\frac{z}{3},\ 2\sqrt{\frac{a}{3}}\cos\left(\frac{2\pi}{3} - \frac{z}{3}\right),\ \text{and}\ 2\sqrt{\frac{a}{3}}\cos\left(\frac{2\pi}{3} + \frac{z}{3}\right);$$

where z is given by the equation $\cos z = \frac{b}{2}\sqrt{\frac{27}{a^3}}$.

EXAMPLES.

1. Solve the equation $x^3 - 4x + 2 = 0$.

Here $a = 4$, $b = -2$; then, $\cos z = -\sqrt{\frac{27}{64}}$, or $\cos(\pi - z) = \sqrt{\frac{27}{64}}$ (§ 33)

By logarithms, $\log\cos(\pi - z) = \frac{1}{2}(\log 27 - \log 64)$.

$$\begin{aligned} \log 27 &= 1.431364 \\ \log 64 &= 1.806180 \\ 2)&\underline{19.625184 - 20} \\ \log\cos(\pi - z) &= 9.812592 - 10 \end{aligned}$$

Then, $\pi - z = 49° 29' 40.5''$, and $z = 130° 30' 19.5''$.

Whence, $\frac{z}{3} = 43° 30' 6.5''$, and $2\sqrt{\frac{a}{3}} = 2\sqrt{\frac{4}{3}} = \sqrt{\frac{16}{3}}$.

Then the three values of x are

$$\sqrt{\frac{16}{3}} \cos 43° 30' 6.5'',$$

$$\sqrt{\frac{16}{3}} \cos (120° - 43° 30' 6.5'') = \sqrt{\frac{16}{3}} \cos 76° 29' 53.5'',$$

and $$\sqrt{\frac{16}{3}} \cos (120° + 43° 30' 6.5'') = \sqrt{\frac{16}{3}} \cos (90° + 73° 30' 6.5'')$$

$$= -\sqrt{\frac{16}{3}} \sin 73° 30' 6.5'' \text{ (§ 30)}.$$

Now,

$$\log \sqrt{\frac{16}{3}} = \tfrac{1}{2}(\log 16 - \log 3) = \tfrac{1}{2}(1.204120 - .477121) = .363500. \quad (1)$$

Also, $$\log \cos 43° 30' \ 6.5'' = 9.860549 - 10, \quad (2)$$

$$\log \cos 76° 29' 53.5'' = 9.368242 - 10, \quad (3)$$

and $$\log \sin 73° 30' \ 6.5'' = 9.981741 - 10. \quad (4)$$

Adding (2), (3), and (4) in succession to (1), the logarithms of the absolute values of x are

$$0.224049, \ 9.731742 - 10, \text{ and } 0.345241.$$

The numbers corresponding to these logarithms are

$$1.67513, \ .53919, \text{ and } 2.21432.$$

Whence, $x = 1.67513, \ .53919, \text{ or } -2.21432.$

Solve the following equations:

2. $x^3 - 4x - 1 = 0.$

3. $x^3 - 6x + 3 = 0.$

4. $x^3 + 6x^2 - x - 1 = 0.$

5. $x^3 - 3x^2 - 2x + 1 = 0.$

SPHERICAL TRIGONOMETRY.

X. GEOMETRICAL PRINCIPLES.

127. If a triedral angle be formed with its vertex at the centre of a sphere, it intercepts on the surface a *spherical triangle.*

The triangle is bounded by three arcs of great circles, called its *sides,* which measure the face angles of the triedral angle.

The *angles* of the spherical triangle are the spherical angles formed by the adjacent sides; and, by Geometry, each is equal to the angle between two straight lines drawn, one in the plane of each of its sides, and perpendicular to the intersection of these planes at the same point.

128. The sides of a spherical triangle are usually expressed in degrees.

129. *Spherical Trigonometry* treats of the trigonometric relations between the sides and angles of a spherical triangle.

The face and diedral angles of the triedral angle are not altered by varying the radius of the sphere; and hence the relations between the sides and angles of a spherical triangle are independent of the length of the radius.

130. We shall limit ourselves in the present work to such triangles as are considered in Geometry, where each angle is less than two right angles, and each side less than the semi-circumference of a great circle; that is, where each element is less than 180°.

131. The proofs of the following properties of spherical triangles may be found in any treatise on Solid Geometry:

1. The sum of any two sides of a spherical triangle is greater than the third side.

2. In any spherical triangle, the greater side lies opposite the greater angle; and, conversely, the greater angle lies opposite the greater side.

3. The sum of the sides of a spherical triangle is less than 360°.

4. The sum of the angles of a spherical triangle is greater than 180°, and less than 540°.

5. If $A'B'C'$ is the polar triangle of ABC, that is, if A, B, and C are the poles of the sides $B'C'$, $C'A'$, and $A'B'$, respectively, then, conversely, ABC is the polar triangle of $A'B'C'$.

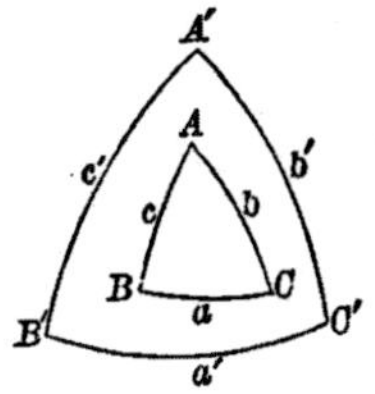

6. In two polar triangles, each angle of one is measured by the supplement of the side lying opposite the homologous angle of the other; that is

$$a' = 180° - A. \qquad b' = 180° - B. \qquad c' = 180° - C.$$

$$A' = 180° - a. \qquad B' = 180° - b. \qquad C' = 180° - c.$$

132. A spherical triangle is called *tri-rectangular* when it has three right angles; each side is a quadrant, and each vertex is the pole of the opposite side.

133. I. Let C be the right angle of the right spherical triangle ABC, and suppose $a < 90°$ and $b < 90°$.

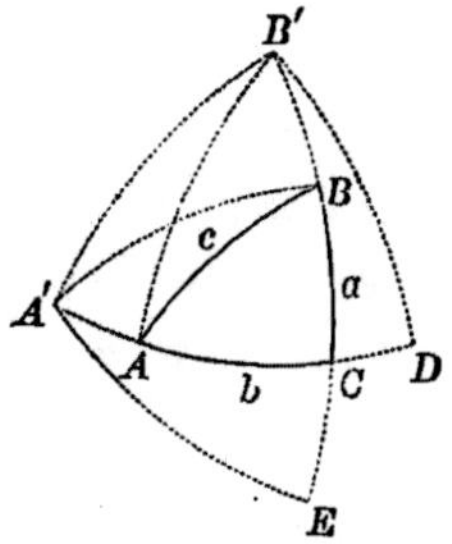

Complete the tri-rectangular triangle $A'B'C$; also, since B' is the pole of AC, and A' of BC, construct the tri-rectangular triangles $AB'D$ and $A'BE$.

Then since B lies within the triangle $AB'D$, AB or c is $< 90°$.

Since BC is $< B'C$, the angle A is $< B'AD$, or $< 90°$.

Since AC is $< A'C$, the angle B is $< A'BE$, or $< 90°$.

II. Suppose $a < 90°$ and $b > 90°$.

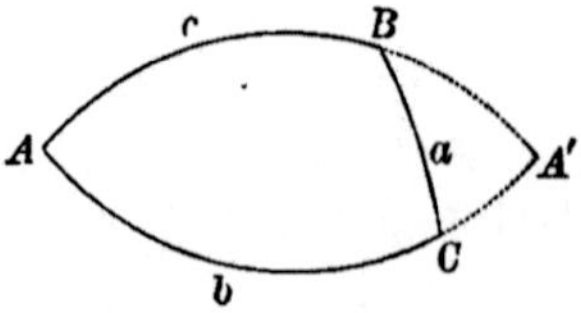

Complete the lune $ABA'C$.

Then in the right triangle $A'BC$, $A'C = 180° - b$.

That is, the sides a and $A'C$ of the triangle $A'BC$ are each $< 90°$; and by I., $A'B$ and the angles A' and $A'BC$ are each $< 90°$.

But, $c = 180° - A'B$, $A = A'$, and $B = 180° - A'BC$.

Whence, c is $> 90°$, $A < 90°$, and $B > 90°$.

Similarly, if a is $> 90°$ and $b < 90°$, then c is $> 90°$, $A > 90°$, and $B < 90°$.

III. Suppose $a > 90°$ and $b > 90°$.

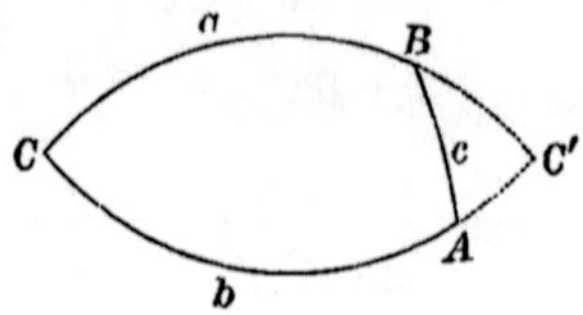

Complete the lune $ACBC'$.

Then in the right triangle ABC', $AC' = 180° - b$, and $BC' = 180° - a$.

That is, the sides AC' and BC' of the triangle ABC' are each $< 90°$; and by I., AB and the angles BAC' and ABC' are each $< 90°$.

But, $A = 180° - BAC'$, and $B = 180° - ABC'$.

Whence, c is $< 90°$, $A > 90°$, and $B > 90°$.

Hence, in any right spherical triangle:

1. *If the sides about the right angle are in the same quadrant, the hypotenuse is* $< 90°$; *if they are in different quadrants, the hypotenuse is* $> 90°$.
2. *An angle is in the same quadrant as its opposite side.*

134. In the figure of § 131, we have, by § 131, 1, $a' < b' + c'$.

Putting for a', b', and c' the values given in § 131, 6, we have

$$180° - A < 180° - B + 180° - C, \text{ or } B + C - A < 180°.$$

Again, by § 130, $B + C + 180° > A$; whence, $B + C - A > -180°$.

Therefore, $B + C - A$ is between $180°$ and $-180°$.

Similarly, $C + A - B$ and $A + B - C$ are between $180°$ and $-180°$.

XI. RIGHT SPHERICAL TRIANGLES.

135. Let C be the right angle of the right spherical triangle ABC.

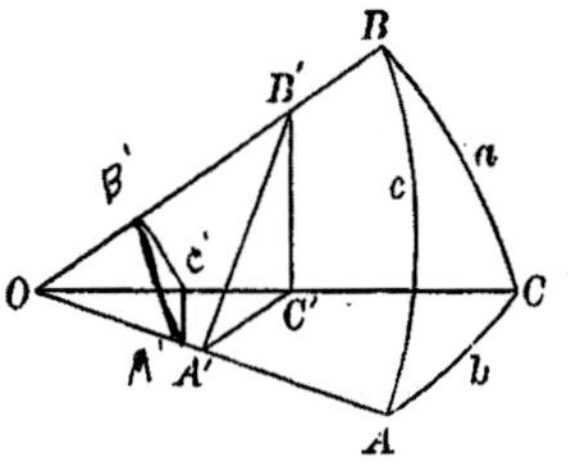

Let O be the centre of the sphere, and draw OA, OB, and OC.

At any point A' of OA draw $A'B'$ and $A'C''$ perpendicular to OA, meeting OB and OC at B' and C'', and draw $B'C''$.

Then OA is perpendicular to the plane $A'B'C''$.

Hence, each of the planes $A'B'C''$ and OBC is perpendicular to the plane OAC, and their intersection $B'C''$ is perpendicular to OAC.

Therefore, $B'C''$ is perpendicular to $A'C''$ and OC''.

In the right triangle $OA'B'$, we have

$$\cos c = \cos A'OB' = \frac{OA'}{OB'} = \frac{OC''}{OB'} \times \frac{OA'}{OC''}.$$

But in the right triangles $OB'C''$ and $OC''A'$,

$$\frac{OC''}{OB'} = \cos a, \text{ and } \frac{OA'}{OC''} = \cos b.$$

Whence, $$\cos c = \cos a \cos b. \quad (75)$$

Again, $$\sin A = \sin B'A'C'' = \frac{B'C''}{A'B'} = \frac{\frac{B'C''}{OB'}}{\frac{A'B'}{OB'}} = \frac{\sin a}{\sin c}. \quad (76)$$

And, $$\cos A = \cos B'A'C'' = \frac{A'C''}{A'B'} = \frac{\frac{A'C''}{OA'}}{\frac{A'B'}{OA'}} = \frac{\tan b}{\tan c}. \quad (77)$$

In like manner, $$\sin B = \frac{\sin b}{\sin c}, \quad (78)$$

and $$\cos B = \frac{\tan a}{\tan c}. \quad (79)$$

136. The proofs of § 135 cannot be regarded as general, for in the construction of the figure we have assumed a and b, and therefore c and A (§ 133), to be less than 90°.

To prove formulæ (**75**) to (**79**) universally, we must consider two additional cases:

CASE I. *When one of the sides a and b is < 90°, and the other > 90°.*

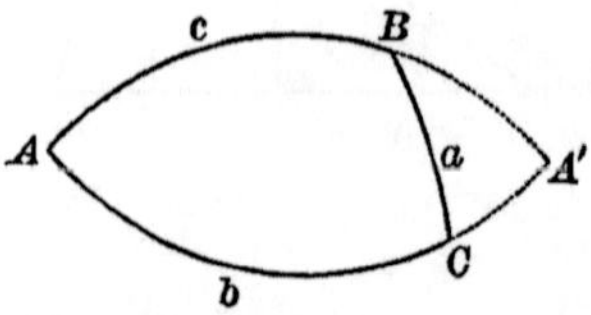

In the right spherical triangle ABC, let a be $< 90°$ and $b > 90°$. Complete the lune $ABA'C$; then, in the spherical triangle $A'BC$,

$$A'B = 180° - c,\ A'C = 180° - b,\ A' = A, \text{ and } A'BC = 180° - B.$$

But by § 133, c is $> 90°$, $A < 90°$, and $B > 90°$.

Hence, each element, except the right angle, of the right spherical triangle $A'BC$ is $< 90°$; and we have by § 135,

$$\cos A'B = \cos a \cos A'C,$$

$$\sin A' = \frac{\sin a}{\sin A'B}, \qquad \sin A'BC = \frac{\sin A'C}{\sin A'B},$$

$$\cos A' = \frac{\tan A'C}{\tan A'B}, \qquad \cos A'BC = \frac{\tan a}{\tan A'B}.$$

Putting for $A'B$, $A'C$, A' and $A'BC$ their values, we have

$$\cos(180° - c) = \cos a \cos(180° - b),$$

$$\sin A = \frac{\sin a}{\sin(180° - c)}, \qquad \sin(180° - B) = \frac{\sin(180° - b)}{\sin(180° - c)},$$

$$\cos A = \frac{\tan(180° - b)}{\tan(180° - c)}, \qquad \cos(180° - B) = \frac{\tan a}{\tan(180° - c)}.$$

Whence, by § 33, $\quad -\cos c = \cos a(-\cos b),$

$$\sin A = \frac{\sin a}{\sin c}, \qquad \sin B = \frac{\sin b}{\sin c},$$

$$\cos A = \frac{-\tan b}{-\tan c}, \qquad -\cos B = \frac{\tan a}{-\tan c};$$

and we obtain formulæ (**75**) to (**79**) as before.

In like manner, the formulæ may be proved to hold when a is $> 90°$ and $b < 90°$.

CASE II. *When both a and b are $> 90°$.*

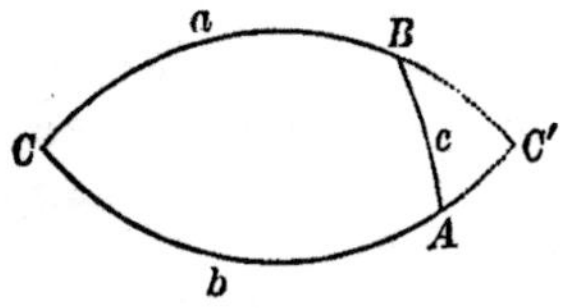

In the right spherical triangle ABC, let a and b be $> 90°$.

Complete the lune $ACBC'$.

By § 133, c is $< 90°$, $A > 90°$, and $B > 90°$.

Hence, each element, except the right angle, of the right spherical triangle ABC' is $< 90°$; and we have by § 135,

$$\cos c = \cos AC' \cos BC',$$

$$\sin BAC' = \frac{\sin BC'}{\sin c}, \qquad \sin ABC' = \frac{\sin AC'}{\sin c},$$

$$\cos BAC' = \frac{\tan AC'}{\tan c}, \qquad \cos ABC' = \frac{\tan BC'}{\tan c}.$$

Putting for AC', BC', BAC', and ABC' their values, we have

$$\cos c = \cos(180° - a)\cos(180° - b),$$

$$\sin(180° - A) = \frac{\sin(180° - a)}{\sin c}, \qquad \sin(180° - B) = \frac{\sin(180° - b)}{\sin c},$$

$$\cos(180° - A) = \frac{\tan(180° - b)}{\tan c}, \qquad \cos(180° - B) = \frac{\tan(180° - a)}{\tan c}.$$

Whence, by § 33, $\quad \cos c = (-\cos a)(-\cos b),$

$$\sin A = \frac{\sin a}{\sin c}, \qquad \sin B = \frac{\sin b}{\sin c},$$

$$-\cos A = \frac{-\tan b}{\tan c}, \qquad -\cos B = \frac{-\tan a}{\tan c};$$

and we obtain formulæ (75) to (79), as before.

137. From (76) and (77), we obtain

$$\tan A = \frac{\sin A}{\cos A} = \frac{\sin a}{\sin c} \times \frac{\tan c}{\tan b} = \frac{\sin a}{\cos c \tan b}.$$

Whence by (75), $\quad \tan A = \dfrac{\sin a}{\cos a \cos b \tan b} = \dfrac{\tan a}{\sin b}.$ (80)

In like manner, $\quad \tan B = \dfrac{\tan b}{\sin a}.$ (81)

138. By (4), $\sin a = \cos a \tan a$; then (76) may be written

$$\sin A = \frac{\cos a \tan a}{\cos c \tan c} = \frac{\frac{\tan a}{\tan c}}{\frac{\cos c}{\cos a}}.$$

Whence by (75) and (79),

$$\sin A = \frac{\cos B}{\cos b}. \quad (82)$$

In like manner, $$\sin B = \frac{\cos A}{\cos a}. \quad (83)$$

139. From (75), (82), and (83), we have

$$\cos c = \cos a \cos b = \frac{\cos A}{\sin B} \times \frac{\cos B}{\sin A} = \cot A \cot B. \quad (84)$$

140. The formulæ of §§ 135 to 139 are collected below for convenience of reference:

$$\cos c = \cos a \cos b.$$

$$\sin A = \frac{\sin a}{\sin c}. \qquad \sin B = \frac{\sin b}{\sin c}.$$

$$\cos A = \frac{\tan b}{\tan c}. \qquad \cos B = \frac{\tan a}{\tan c}.$$

$$\tan A = \frac{\tan a}{\sin b}. \qquad \tan B = \frac{\tan b}{\sin a}.$$

$$\sin A = \frac{\cos B}{\cos b}. \qquad \sin B = \frac{\cos A}{\cos a}.$$

$$\cos c = \cot A \cot B.$$

The student should compare the formulæ for the sines, cosines, and tangents of A and B with the corresponding formulæ in §§ 2 and 5.

141. Napier's Rules of Circular Parts.

These are two rules which include all the formulæ of § 140.

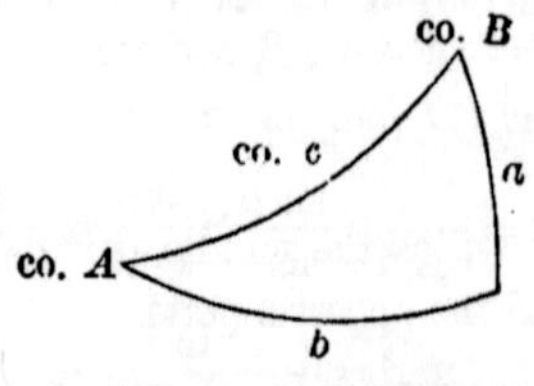

In any right spherical triangle, the elements a and b, and the complements of the elements A, B, and c (written in abbreviated form, co. A, co. B, and co. c), are called the *circular parts.*

If we suppose them arranged in the order in which the letters occur in the triangle, any one of the five may be taken and called the *middle part*; the two immediately adjacent are called the *adjacent parts*, and the remaining two the *opposite parts*.

Then Napier's rules are:

I. *The sine of the middle part is equal to the product of the tangents of the adjacent parts.*

II. *The sine of the middle part is equal to the product of the cosines of the opposite parts.*

142. Napier's rules may be proved by taking each circular part in succession as the middle part, and showing that the results agree with the formulæ of § 140.

1. If a be taken as the middle part, b and co. B are the adjacent parts, and co. c and co. A the opposite parts.

Then the rules give

$$\sin a = \tan b \tan(\text{co. } B), \text{ and } \sin a = \cos(\text{co. } c)\cos(\text{co. } A).$$

Or by § 32, $\sin a = \tan b \cot B$, and $\sin a = \sin c \sin A$;

which are equivalent to (**81**) and (**76**).

2. If b be taken as the middle part, a and co. A are the adjacent parts, and co. c and co. B the opposite parts.

Then, $$\sin b = \tan a \tan(\text{co. } A) = \tan a \cot A,$$

and $$\sin b = \cos(\text{co. } c)\cos(\text{co. } B) = \sin c \sin B;$$

which are equivalent to (**80**) and (**78**).

3. If co. c be taken as the middle part, co. A and co. B are the adjacent parts, and a and b the opposite parts.

Then, $\sin(\text{co. } c) = \tan(\text{co. } A)\tan(\text{co. } B)$, and $\sin(\text{co. } c) = \cos a \cos b$.

Or, $$\cos c = \cot A \cot B, \text{ and } \cos c = \cos a \cos b;$$

which agree with (**84**) and (**75**).

4. If co. A be taken as the middle part, b and co. c are the adjacent parts, and a and co. B the opposite parts.

Then, $\sin(\text{co. } A) = \tan b \tan(\text{co. } c)$, and $\sin(\text{co. } A) = \cos a \cos(\text{co. } B)$.

Or, $$\cos A = \tan b \cot c, \text{ and } \cos A = \cos a \sin B;$$

which are equivalent to (**77**) and (**83**).

5. If co. B be taken as the middle part, a and co. c are the adjacent parts, and b and co. A the opposite parts.

Then, $\sin(\text{co. } B) = \tan a \tan(\text{co. } c)$, and $\sin(\text{co. } B) = \cos b \cos(\text{co. } A)$.

Or, $\cos B = \tan a \cot c$, and $\cos B = \cos b \sin A$;

which are equivalent to **(79)** and **(82)**.

Writers on Trigonometry differ as to the practical value of Napier's rules; but in the opinion of the highest authorities, it seems to be regarded as preferable to attempt to remember the formulæ by comparing them with the analogous formulæ for plane right triangles, as stated in § 140.

SOLUTION OF RIGHT SPHERICAL TRIANGLES.

143. To solve a right spherical triangle, two elements must be given in addition to the right angle.

There may be six cases:

1. *Given the hypotenuse and an adjacent angle.*
2. *Given an angle and its opposite side.*
3. *Given an angle and its adjacent side.*
4. *Given the hypotenuse and another side.*
5. *Given the two sides a and b.*
6. *Given the two angles A and B.*

144. Either of the above cases may be solved by aid of § 140.

The formula for computing either of the remaining elements when any two are given may be found by the following rule:

Take that formula which involves the given parts and the required part.

If *all* the remaining elements are required, the following rule may be found convenient in selecting the formulæ:

Take the three formulæ which involve the given parts.

145. It is convenient in the solution to have a check on the logarithmic work, which may be done in every case without the necessity of looking out any new logarithms.

Examples of this will be found in § 148.

The check formula for any particular case may be selected from the set in § 140 by the following rule:

Take that formula which involves the three required parts.

Note. If Napier's rules are used, the following rule will indicate which of the circular parts corresponding to the given elements and any required element is to be regarded as the middle part.

If these three circular parts are adjacent, take the middle one as the middle part, and the others are then adjacent parts.

If they are not adjacent, take the part which is not adjacent to either of the others as the middle part, and the others are then opposite parts.

For the check formula, proceed as above with the circular parts corresponding to the three required elements.

Thus, if c and A are the given elements,

1. To find a, consider the circular parts a, co. c, and co. A; of these, a is the middle part, and co. c and co. A are opposite parts. Then, by Napier's rules,

$$\sin a = \cos(\text{co. } c)\cos(\text{co. } A) = \sin c \sin A.$$

2. To find b, the circular parts are b, co. c, and co. A; in this case co. A is the middle part, and b and co. c are adjacent parts. Then,

$$\sin(\text{co. } A) = \tan b \tan(\text{co. } c), \text{ or } \cos A = \tan b \cot c.$$

3. To find B, the circular parts are co. B, co. c, and co. A; co. c is the middle part, and co. A and co. B are adjacent parts. Then,

$$\sin(\text{co. } c) = \tan(\text{co. } A)\tan(\text{co. } B), \text{ or } \cos c = \cot A \cot B.$$

4. For the check formula, the circular parts are a, b, and co. B; a is the middle part, and b and co. B are adjacent parts. Then,

$$\sin a = \tan b \tan(\text{co. } B) = \tan b \cot B.$$

146. In solving spherical triangles, careful attention must be given to the *algebraic signs* of the functions; the cosines, tangents, and cotangents of angles between 90° and 180° being taken *negative* (§ 20).

It is convenient to place the sign of each function just above or below it, as shown in the examples of § 148; the sign of the function in the first member being then determined in accordance with the principle that like signs produce +, and unlike signs produce −.

Note. In the examples after the first of § 148, the signs are omitted in every case where both factors of the second member are +.

147. In finding the angles corresponding, if the function is a cosine, tangent, or cotangent, its sign determines whether the angle is acute or obtuse; that is, if it is +, the angle is acute; and if it is −, the angle is obtuse, and the *supplement* of the acute angle obtained from the tables must be taken (§ 33).

If the function is a sine, since the sine of an angle is equal to the sine of its supplement (§ 33), both the acute angle obtained from the tables and its supplement must be retained as solutions, unless the ambiguity can be removed by the principles of § 133.

EXAMPLES.

148. 1. Given $B = 33° 50'$, $a = 108°$; find A, b, and c.

By the rule of § 144, the formulæ from § 140 are,

$$\sin B = \frac{\cos A}{\cos a},\quad \tan B = \frac{\tan b}{\sin a},\quad \text{and}\quad \cos B = \frac{\tan a}{\tan c}.$$

$$\text{Or,}\quad \overset{-}{\cos A} = \overset{-}{\cos a}\,\overset{+}{\sin B},\quad \overset{+}{\tan b} = \overset{+}{\sin a}\,\overset{+}{\tan B},\quad \text{and}\quad \overset{-}{\tan c} = \frac{\overset{+}{\tan a}}{\underset{+}{\cos B}}.$$

Hence,

$$\log\cos A = \log\cos a + \log\sin B.$$
$$\log\tan b = \log\sin a + \log\tan B.$$
$$\log\tan c = \log\tan a - \log\cos B.$$

Since $\cos A$ and $\tan c$ are negative, the *supplements* of the acute angles obtained from the tables must be taken (§ 147).

Note 1. When the supplement of the angle obtained from the tables is to be taken, it is convenient to write 180° minus the element in the first member, as shown below in the cases of A and c.

By the rule of § 145, the check formula for this case is

$$\cos A = \frac{\tan b}{\tan c},\quad \text{or}\quad \log\cos A = \log\tan b - \log\tan c.$$

The values of $\log\tan b$ and $\log\tan c$ may be taken from the first part of the work, and their difference should be equal to the result previously found for $\log\cos A$.

$\log\cos a = 9.489982 - 10$	$\log\tan a = 0.488224$
$\log\sin B = 9.745683 - 10$	$\log\cos B = 9.919424 - 10$
$\log\cos A = 9.235665 - 10$	$\log\tan c = 0.568800$
$180° - A = 80° 5' 33.8''.$	$180° - c = 74° 53' 45.0''.$
$A = 99° 54' 26.2''.$	$c = 105° 6' 15.0''.$
$\log\sin a = 9.978206 - 10$	*Check.*
$\log\tan B = 9.826259 - 10$	$\log\tan b = 9.804465 - 10$
$\log\tan b = 9.804465 - 10$	$\log\tan c = 0.568800$
$b = 32° 30' 59.8''.$	$\log\cos A = 9.235665 - 10$

2. Given $c = 70° 30'$, $A = 100°$; find a, b, and B.

In this case the three formulæ are

$$\sin A = \frac{\sin a}{\sin c},\quad \cos A = \frac{\tan b}{\tan c},\quad \text{and}\quad \cos c = \cot A \cot B.$$

$$\text{Or,}\quad \sin a = \sin c \sin A,\quad \overset{-}{\tan b} = \overset{+}{\tan c}\,\overset{-}{\cos A},\quad \text{and}\quad \overset{-}{\cot B} = \overset{+}{\cos c}\,\overset{-}{\tan A}.$$

Here the side a is determined from its sine; but the ambiguity is removed by the principles of § 133; for a and A must be in the same quadrant. Therefore, a is obtuse; and the supplement of the angle obtained from the tables must be taken.

By § 145, the check formula is

$$\tan B = \frac{\tan b}{\sin a}, \text{ or } \sin a = \tan b \cot B.$$

Note 2. The check formula should always be expressed in terms of the functions used in determining the required parts; thus, in the case above, the check formula is transformed so as to involve $\cot B$ instead of $\tan B$.

$$\begin{aligned} \log\sin c &= 9.974347-10 \\ \log\sin A &= 9.993351-10 \\ \hline \log\sin a &= 9.967698-10 \\ 180^\circ - a &= 68^\circ 10' 28.2''. \\ a &= 111^\circ 49' 31.8''. \end{aligned}$$

$$\begin{aligned} \log\cos c &= 9.523495-10 \\ \log\tan A &= 0.753681 \\ \hline \log\cot B &= 0.277176 \\ 180^\circ - B &= 27^\circ 50' 39.8''. \\ B &= 152^\circ 9' 20.2''. \end{aligned}$$

$$\begin{aligned} \log\tan c &= 0.450851 \\ \log\cos A &= 9.239670-10 \\ \hline \log\tan b &= 9.690521-10 \\ 180^\circ - b &= 26^\circ 7' 18.4''. \\ b &= 153^\circ 52' 41.6''. \end{aligned}$$

Check.

$$\begin{aligned} \log\tan b &= 9.690521-10 \\ \log\cot B &= 0.277176 \\ \hline \log\sin a &= 9.967697-10 \end{aligned}$$

Note 3. We observe here a difference of .000001 in the two values of $\log\sin a$. This does not necessarily indicate an error in the work, for such a small difference might easily be due to the fact that the logarithms are only *approximately* correct to the sixth decimal place.

3. Given $a = 132^\circ 6'$, $b = 77^\circ 51'$; find A, B, and c.

In this case the three formulæ are

$$\overset{-}{\tan} A = \frac{\overset{-}{\tan a}}{\underset{+}{\sin b}}, \quad \tan B = \frac{\tan b}{\sin a}, \text{ and } \overset{-}{\cos} c = \overset{-}{\cos} a \overset{+}{\cos} b.$$

The check formula is

$$\cos c = \cot A \cot B, \text{ or } \cos c \tan A \tan B = 1.$$

That is, $\log\cos c + \log\tan A + \log\tan B = \log 1 = 0.$

$$\begin{aligned} \log\tan a &= 0.044039 \\ \log\sin b &= 9.990161-10 \\ \hline \log\tan A &= 0.053878 \\ 180^\circ - A &= 48^\circ 32' 41.8''. \\ A &= 131^\circ 27' 18.2''. \end{aligned}$$

$$\begin{aligned} \log\cos a &= 9.826351-10 \\ \log\cos b &= 9.323194-10 \\ \hline \log\cos c &= 9.149545-10 \\ 180^\circ - c &= 81^\circ 53' 17.4''. \\ c &= 98^\circ 6' 42.6''. \end{aligned}$$

	Check.
$\log\tan b = 0.666967$	$\log\cos c = 9.149545 - 10$
$\log\sin a = 9.870390 - 10$	$\log\tan A = 0.053878$
$\log\tan B = 0.796577$	$\log\tan B = 0.796577$
$B = 80°\,55'\,26.6''.$	$\log 1 = 0.000000$

4. Given $A = 105°\,59'$, $a = 128°\,33'$; find b, B, and c.

The formulæ are

$$\overset{+}{\sin} b = \frac{\overset{-}{\tan a}}{\underset{-}{\tan A}},\quad \overset{+}{\sin} B = \frac{\overset{-}{\cos A}}{\underset{-}{\cos a}},\quad \text{and}\ \sin c = \frac{\sin a}{\sin A}.$$

The check formula is $\sin B = \dfrac{\sin b}{\sin c}$.

In this case, each of the required parts is determined from its sine; and as the ambiguity cannot be removed by § 133, both the acute angle obtained from the tables and its supplement must be retained in each case.

$\log\tan a = 0.098617$	$\log\sin a = 9.893243 - 10$
$\log\tan A = 0.542981$	$\log\sin A = 9.982878 - 10$
$\log\sin b = 9.555636 - 10$	$\log\sin c = 9.910365 - 10$
$b = 21°\,3'\,58.7''$	$c = 54°\,26'\,26.7''$
or $158°\,56'\,1.3''$.	or $125°\,33'\,33.3''$.

	Check.
$\log\cos A = 9.439897 - 10$	$\log\sin b = 9.555636 - 10$
$\log\cos a = 9.794626 - 10$	$\log\sin c = 9.910365 - 10$
$\log\sin B = 9.645271 - 10$	$\log\sin B = 9.645271 - 10$
$B = 26°\,13'\,18.2''$	
or $153°\,46'\,41.8''$.	

It does not follow, however, that these values can be combined promiscuously; for by § 133, since a is $> 90°$, with the value of b less than $90°$ must be taken the value of c greater than $90°$, and the value of B less than $90°$; while with the value of b greater than $90°$ must be taken the value of c less than $90°$, and the value of B greater than $90°$.

Thus the only solutions of the example are:

1. $b = 21°\,3'\,58.7''$, $c = 125°\,33'\,33.3''$, $B = 26°\,13'\,18.2''$.

2. $b = 158°\,56'\,1.3''$, $c = 54°\,26'\,26.7''$, $B = 153°\,46'\,41.8''$.

Note 4. The figure shows geometrically why there are two solutions in this case.

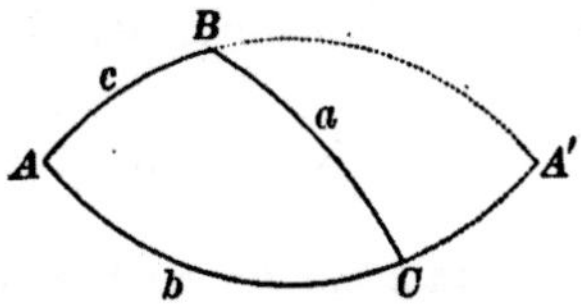

For if AB and AC be produced to A', forming the lune $ABA'C$, the triangle $A'BC$ has the side a and the angle A' equal, respectively, to the side a and the angle A of the triangle ABC, and both triangles are right-angled at C.

It is evident that the sides $A'B$ and $A'C$ and the angle $A'BC$ are the supplements of the sides c and b and the angle ABC, respectively.

Solve the following right spherical triangles:

5. Given $c = 49°$, $a = 27°$.
6. Given $A = 38°$, $B = 63°$.
7. Given $A = 31°$, $a = 23°$.
8. Given $B = 153°$, $a = 35°$.
9. Given $a = 15°$, $b = 106°$.
10. Given $c = 139°$, $A = 165°$.
11. Given $B = 82° 25'$, $b = 68° 35'$.
12. Given $c = 75° 37'$, $B = 29° 4'$.
13. Given $c = 118° 49'$, $b = 44° 23'$.
14. Given $a = 171° 6'$, $b = 161° 58'$.
15. Given $B = 100° 40'$, $a = 170° 38'$.
16. Given $A = 102° 57'$, $B = 143° 46'$.
17. Given $a = 10° 28'$, $b = 7° 10'$.
18. Given $A = 54° 11'$, $b = 83° 29'$.
19. Given $A = 50° 43'$, $B = 122° 18'$.
20. Given $c = 59° 3'$, $A = 147° 32'$.
21. Given $B = 103° 30'$, $b = 132° 54'$.
22. Given $A = 95° 15'$, $b = 166° 7'$.
23. Given $c = 78° 52'$, $a = 114° 26'$.
24. Given $c = 127° 9'$, $B = 80° 51'$.
25. Given $A = 98° 34'$, $a = 113° 12'$.
26. Given $c = 136° 21'$, $b = 157° 41'$.

149. Quadrantal Triangles.

A spherical triangle is called *quadrantal* when it has one side equal to a quadrant.

By § 131, 6, the polar triangle of a quadrantal triangle is a *right* spherical triangle.

Hence, to solve a quadrantal triangle, we have only to solve its polar triangle, and take the *supplements* of the results.

1. Given $c = 90°$, $a = 67° 38'$, $b = 48° 50'$; find A, B, and C.

Denoting the polar triangle by $A'B'C'$, we have by § 131, 6:

$$C' = 90°,\ A' = 112° 22',\ B' = 131° 10';\ \text{to find } a', b', \text{ and } c'.$$

By § 144, the formulæ for the solution are

$$\overset{-}{\cos a'} = \frac{\overset{-}{\cos A'}}{\underset{+}{\sin B'}},\quad \overset{-}{\cos b'} = \frac{\overset{-}{\cos B'}}{\underset{+}{\sin A'}},\quad \text{and}\quad \overset{+}{\cos c'} = \overset{-}{\cot A'}\,\overset{-}{\cot B'}.$$

The check formula is $\cos c' = \cos a' \cos b'$.

$$\begin{aligned} \log \cos A' &= 9.580392 - 10 \\ \log \sin B' &= 9.876678 - 10 \\ \hline \log \cos a' &= 9.703714 - 10 \\ 180° - a' &= 59° 38' 9.7''. \end{aligned} \qquad \begin{aligned} \log \cot A' &= 9.614359 - 10 \\ \log \cot B' &= 9.941713 - 10 \\ \hline \log \cos c' &= 9.556072 - 10 \\ c' &= 68° 54' 41.5''. \end{aligned}$$

$$\begin{aligned} \log \cos B' &= 9.818392 - 10 \\ \log \sin A' &= 9.966033 - 10 \\ \hline \log \cos b' &= 9.852359 - 10 \\ 180° - b' &= 44° 37' 5.8''. \end{aligned} \qquad \begin{aligned} &\textit{Check.} \\ \log \cos a' &= 9.703714 - 10 \\ \log \cos b' &= 9.852359 - 10 \\ \hline \log \cos c' &= 9.556073 - 10 \end{aligned}$$

Then in the given quadrantal triangle, we have

$$A = 180° - a' = 59° 38' 9.7'',$$
$$B = 180° - b' = 44° 37' 5.8'',$$
$$C = 180° - c' = 111° 5' 18.5''.$$

EXAMPLES.

Solve the following quadrantal triangles:

2. Given $A = 122°$, $b = 154°$.
3. Given $A = 45° 52'$, $B = 139° 24'$.
4. Given $a = 30° 19'$, $C = 42° 31'$.
5. Given $B = 51° 35'$, $C = 116° 13'$.
6. Given $A = 105° 8'$, $a = 104° 56'$.
7. Given $a = 67° 27'$, $b = 81° 40'$.

150. Isosceles Spherical Triangles.

We know, by Geometry, that if an arc of a great circle be drawn from the vertex of an isosceles spherical triangle to the middle point of the base, it is perpendicular to the base, bisects the vertical angle, and divides the triangle into two symmetrical right spherical triangles.

By solving one of these, we can find the required parts of the given triangle.

1. Given $a=115°$, $b=115°$, $C=71°\,48'$; find A, B, and c.

Denoting the elements of one of the right triangles by A', B', C', a', b', and c', where C' is the right angle, we have

$$c'=a=115°, \text{ and } A'=\tfrac{1}{2}C=35°\,54'.$$

We have then to find the parts a' and B' in this triangle.

By § 140, $\sin A'=\dfrac{\sin a'}{\sin c'}$, and $\cos c'=\cot A'\cot B'$.

Or, $\sin a'=\sin c'\sin A'$, and $\overset{-}{\cot} B'=\overset{-}{\cos} c'\overset{+}{\tan} A'$.

$$\begin{aligned}
\log\sin c' &= 9.957276-10 \\
\log\sin A' &= 9.768173-10 \\
\hline
\log\sin a' &= 9.725449-10 \\
a' &= 32°\,6'\,8.6''.
\end{aligned}
\qquad
\begin{aligned}
\log\cos c' &= 9.625948-10 \\
\log\tan A' &= 9.859666-10 \\
\hline
\log\cot B' &= 9.485614-10 \\
180°-B' &= 72°\,59'\,23.5''. \\
B' &= 107°\,0'\,36.5''.
\end{aligned}$$

Then in the given isosceles triangle,

$$A=B=B'=107°\,0'\,36.5'', \text{ and } c=2a'=64°\,12'\,17.2''.$$

EXAMPLES.

Solve the following isosceles spherical triangles:

2. Given $A=27°\,12'$, $B=27°\,12'$, $c=135°\,20'$.
3. Given $a=152°\,6'$, $b=152°\,6'$, $C=67°\,46'$.
4. Given $a=112°\,25'$, $b=112°\,25'$, $c=123°\,48'$.
5. Given $A=159°$, $B=159°$, $a=137°\,39'$.

XII. OBLIQUE SPHERICAL TRIANGLES.

GENERAL PROPERTIES OF SPHERICAL TRIANGLES.

151. *In any spherical triangle, the sines of the sides are proportional to the sines of their opposite angles.*

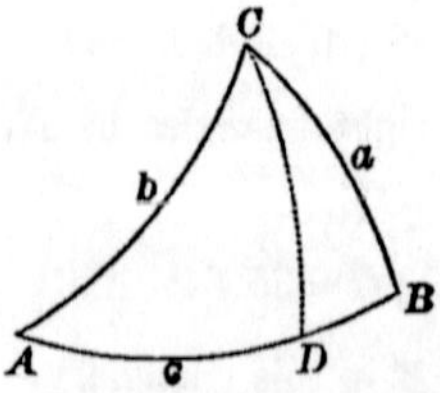

Fig. 1.

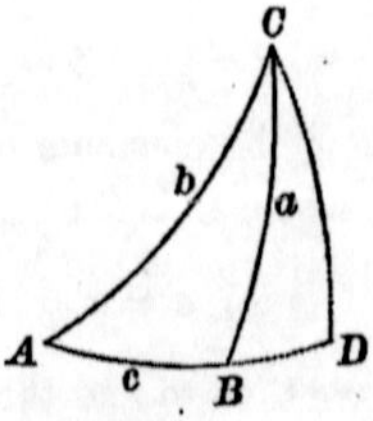

Fig. 2.

Let ABC be any spherical triangle, and draw the arc CD perpendicular to AB.

There will be two cases according as CD falls upon AB (Fig. 1), or upon AB produced (Fig. 2).

In the right triangle ACD, in either figure, we have

$$\sin A = \frac{\sin CD}{\sin b}, \text{ by } (76).$$

Also, in Fig. 1, $$\sin B = \frac{\sin CD}{\sin a}.$$

And in Fig. 2, $$\sin B = \sin (180^\circ - CBD)$$
$$= \sin CBD \ (\S\ 33) = \frac{\sin CD}{\sin a}.$$

Dividing these equations, we have in either case

$$\frac{\sin A}{\sin B} = \frac{\dfrac{\sin CD}{\sin b}}{\dfrac{\sin CD}{\sin a}} = \frac{\sin a}{\sin b}. \tag{85}$$

In like manner, $$\frac{\sin B}{\sin C} = \frac{\sin b}{\sin c}, \tag{86}$$

and $$\frac{\sin A}{\sin C} = \frac{\sin a}{\sin c}. \tag{87}$$

152. *In any spherical triangle, the cosine of any side is equal to the product of the cosines of the other two sides, plus the continued product of their sines and the cosine of their included angle.*

In the right triangle BCD, in Fig. 1, § 151, we have, by (**75**),

$$\cos a = \cos BD \cos CD = \cos (c - AD) \cos CD.$$

And in Fig. 2,

$$\cos a = \cos BD \cos CD = \cos (AD - c) \cos CD.$$

Whence, in either case, by (**12**),

$$\cos a = \cos c \cos AD \cos CD + \sin c \sin AD \cos CD.$$

But in the right triangle ACD,

$$\cos AD \cos CD = \cos b, \text{ by } (\mathbf{75}).$$

And, $\sin AD \cos CD = \sin AD \dfrac{\cos b}{\cos AD} = \cos b \tan AD$

$$= \sin b \frac{\tan AD}{\tan b} = \sin b \cos A, \text{ by } (\mathbf{77}).$$

Whence, $\cos a = \cos b \cos c + \sin b \sin c \cos A.$ (**88**)

In like manner, $\cos b = \cos c \cos a + \sin c \sin a \cos B,$ (**89**)

and $\cos c = \cos a \cos b + \sin a \sin b \cos C.$ (**90**)

153. Let ABC and $A'B'C'$ be a pair of polar triangles.

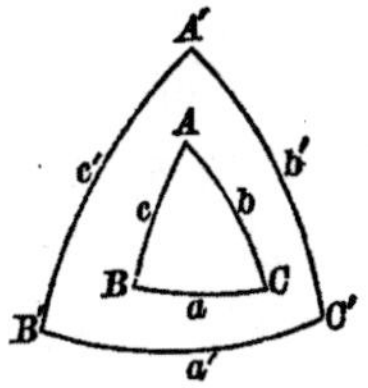

Applying formula (**88**) to the triangle $A'B'C'$, we obtain

$$\cos a' = \cos b' \cos c' + \sin b' \sin c' \cos A'.$$

Putting for a', b', c', and A' the values given in § 131, 6, we have

$$\cos (180° - A) = \cos (180° - B) \cos (180° - C)$$
$$+ \sin (180° - B) \sin (180° - C) \cos (180° - a).$$

Whence, $-\cos A = (-\cos B)(-\cos C) + \sin B \sin C (-\cos a)$ (§ 33).

That is, $\cos A = -\cos B \cos C + \sin B \sin C \cos a.$ (**91**)

Similarly, $\cos B = -\cos C \cos A + \sin C \sin A \cos b,$ (**92**)

and $\cos C = -\cos A \cos B + \sin A \sin B \cos c.$ (**93**)

The above proof illustrates a very important application of the theory of polar triangles to Spherical Trigonometry.

If any relation has been found between the elements of a spherical triangle, an analogous relation may be derived from it, in which each side or angle is replaced by the opposite angle or side, with suitable modifications in the algebraic signs.

154. *To express the sines, cosines, and tangents of the half-angles of a spherical triangle in terms of the sides of the triangle.*

From (88), § 152 $\sin b \sin c \cos A = \cos a - \cos b \cos c.$

Whence, $$\cos A = \frac{\cos a - \cos b \cos c}{\sin b \sin c}. \qquad (A)$$

Subtracting both members from 1, we have

$$1 - \cos A = 1 - \frac{\cos a - \cos b \cos c}{\sin b \sin c} = \frac{\cos b \cos c + \sin b \sin c - \cos a}{\sin b \sin c}.$$

Whence, by (31), $$2 \sin^2 \tfrac{1}{2} A = \frac{\cos (b - c) - \cos a}{\sin b \sin c}.$$

But by (20), $$\cos y - \cos x = 2 \sin \tfrac{1}{2} (x + y) \sin \tfrac{1}{2} (x - y). \qquad (B)$$

Whence, $$2 \sin^2 \tfrac{1}{2} A = \frac{2 \sin \tfrac{1}{2} [a + (b - c)] \sin \tfrac{1}{2} [a - (b - c)]}{\sin b \sin c},$$

or, $$\sin^2 \tfrac{1}{2} A = \frac{\sin \tfrac{1}{2} (a + b - c) \sin \tfrac{1}{2} (a - b + c)}{\sin b \sin c}.$$

Denoting the sum of the sides, $a + b + c$, by $2s$, we have

$$a + b - c = (a + b + c) - 2c = 2s - 2c = 2(s - c),$$

and $$a - b + c = (a + b + c) - 2b = 2s - 2b = 2(s - b).$$

Whence, $$\sin^2 \tfrac{1}{2} A = \frac{\sin (s - b) \sin (s - c)}{\sin b \sin c}.$$

Or, $$\sin \tfrac{1}{2} A = \sqrt{\frac{\sin (s - b) \sin (s - c)}{\sin b \sin c}}. \qquad (94)$$

In like manner, $$\sin \tfrac{1}{2} B = \sqrt{\frac{\sin (s - c) \sin (s - a)}{\sin c \sin a}}, \qquad (95)$$

and $$\sin \tfrac{1}{2} C = \sqrt{\frac{\sin (s - a) \sin (s - b)}{\sin a \sin b}}. \qquad (96)$$

Again, adding both members of (A) to 1, we have

$$1 + \cos A = 1 + \frac{\cos a - \cos b \cos c}{\sin b \sin c} = \frac{\cos a - (\cos b \cos c - \sin b \sin c)}{\sin b \sin c}.$$

Whence, by (32), $$2 \cos^2 \tfrac{1}{2} A = \frac{\cos a - \cos (b + c)}{\sin b \sin c}.$$

Then by (B), $2\cos^2\frac{1}{2}A = \frac{2\sin\frac{1}{2}(b+c+a)\sin\frac{1}{2}(b+c-a)}{\sin b\sin c}.$

Putting $a+b+c=2s$, whence $b+c-a=2(s-a)$, we have

$$\cos^2\tfrac{1}{2}A = \frac{\sin s\sin(s-a)}{\sin b\sin c}.$$

Or, $$\cos\tfrac{1}{2}A = \sqrt{\frac{\sin s\sin(s-a)}{\sin b\sin c}}. \qquad (97)$$

In like manner, $$\cos\tfrac{1}{2}B = \sqrt{\frac{\sin s\sin(s-b)}{\sin c\sin a}}, \qquad (98)$$

and $$\cos\tfrac{1}{2}C = \sqrt{\frac{\sin s\sin(s-c)}{\sin a\sin b}}. \qquad (99)$$

Dividing (94) by (97), we have

$$\tan\tfrac{1}{2}A = \sqrt{\frac{\sin(s-b)\sin(s-c)}{\sin b\sin c}}\sqrt{\frac{\sin b\sin c}{\sin s\sin(s-a)}}$$

$$= \sqrt{\frac{\sin(s-b)\sin(s-c)}{\sin s\sin(s-a)}}. \qquad (100)$$

In like manner, $$\tan\tfrac{1}{2}B = \sqrt{\frac{\sin(s-c)\sin(s-a)}{\sin s\sin(s-b)}}, \qquad (101)$$

and $$\tan\tfrac{1}{2}C = \sqrt{\frac{\sin(s-a)\sin(s-b)}{\sin s\sin(s-c)}}. \qquad (102)$$

155. *To express the sines, cosines, and tangents of the half-sides of a spherical triangle in terms of the angles of the triangle.*

From (91), § 153, $\sin B\sin C\cos a = \cos A + \cos B\cos C$.

Whence, $$\cos a = \frac{\cos A + \cos B\cos C}{\sin B\sin C}. \qquad (A)$$

Then, $$1-\cos a = 1 - \frac{\cos A + \cos B\cos C}{\sin B\sin C}.$$

Or, $$2\sin^2\tfrac{1}{2}a = \frac{-(\cos B\cos C - \sin B\sin C) - \cos A}{\sin B\sin C}$$

$$= -\frac{\cos(B+C) + \cos A}{\sin B\sin C}.$$

Then by (19), $$2\sin^2\tfrac{1}{2}a = -\frac{2\cos\frac{1}{2}(B+C+A)\cos\frac{1}{2}(B+C-A)}{\sin B\sin C}.$$

Denoting the sum of the angles, $A+B+C$, by $2S$, we have

$$B+C-A=2(S-A).$$

Whence, $$\sin^2\tfrac{1}{2}a=-\frac{\cos S\cos(S-A)}{\sin B\sin C}.$$

Or, $$\sin\tfrac{1}{2}a=\sqrt{-\frac{\cos S\cos(S-A)}{\sin B\sin C}}. \qquad (103)$$

In like mannner, $$\sin\tfrac{1}{2}b=\sqrt{-\frac{\cos S\cos(S-B)}{\sin C\sin A}}, \qquad (104)$$

and $$\sin\tfrac{1}{2}c=\sqrt{-\frac{\cos S\cos(S-C)}{\sin A\sin B}}. \qquad (105)$$

Again, adding both members of (A) to 1, we have

$$1+\cos a=1+\frac{\cos A+\cos B\cos C}{\sin B\sin C}=\frac{\cos A+\cos B\cos C+\sin B\sin C}{\sin B\sin C}.$$

Then, $$2\cos^2\tfrac{1}{2}a=\frac{\cos A+\cos(B-C)}{\sin B\sin C}$$

$$=\frac{2\cos\frac{1}{2}[A+B-C]\cos\frac{1}{2}[A-(B-C)]}{\sin B\sin C}.$$

Or, $$\cos^2\tfrac{1}{2}a=\frac{\cos\frac{1}{2}(A+B-C)\cos\frac{1}{2}(A-B+C)}{\sin B\sin C}.$$

But $A+B-C=2(S-C)$, and $A-B+C=2(S-B)$.

Whence, $$\cos^2\tfrac{1}{2}a=\frac{\cos(S-B)\cos(S-C)}{\sin B\sin C}.$$

Or, $$\cos\tfrac{1}{2}a=\sqrt{\frac{\cos(S-B)\cos(S-C)}{\sin B\sin C}}. \qquad (106)$$

In like manner, $$\cos\tfrac{1}{2}b=\sqrt{\frac{\cos(S-C)\cos(S-A)}{\sin C\sin A}}, \qquad (107)$$

and $$\cos\tfrac{1}{2}c=\sqrt{\frac{\cos(S-A)\cos(S-B)}{\sin A\sin B}}. \qquad (108)$$

Dividing (**103**) by (**106**), we have

$$\tan\tfrac{1}{2}a=\sqrt{-\frac{\cos S\cos(S-A)}{\cos(S-B)\cos(S-C)}}. \qquad (109)$$

In like manner, $$\tan\tfrac{1}{2}b=\sqrt{-\frac{\cos S\cos(S-B)}{\cos(S-C)\cos(S-A)}}, \qquad (110)$$

and $$\tan\tfrac{1}{2}c=\sqrt{-\frac{\cos S\cos(S-C)}{\cos(S-A)\cos(S-B)}}. \qquad (111)$$

NAPIER'S ANALOGIES.

156. Dividing (**100**) by (**101**), we have

$$\frac{\tan\frac{1}{2}A}{\tan\frac{1}{2}B}=\sqrt{\frac{\sin(s-b)\sin(s-c)}{\sin s\sin(s-a)}}\sqrt{\frac{\sin s\sin(s-b)}{\sin(s-c)\sin(s-a)}}$$

Or,
$$\frac{\sin\frac{1}{2}A\cos\frac{1}{2}B}{\cos\frac{1}{2}A\sin\frac{1}{2}B}=\sqrt{\frac{\sin^2(s-b)}{\sin^2(s-a)}}=\frac{\sin(s-b)}{\sin(s-a)}.$$

Whence by composition and division,

$$\frac{\sin\frac{1}{2}A\cos\frac{1}{2}B+\cos\frac{1}{2}A\sin\frac{1}{2}B}{\sin\frac{1}{2}A\cos\frac{1}{2}B-\cos\frac{1}{2}A\sin\frac{1}{2}B}=\frac{\sin(s-b)+\sin(s-a)}{\sin(s-b)-\sin(s-a)}.$$

Then by (**9**), (**11**), and (**21**),

$$\frac{\sin(\frac{1}{2}A+\frac{1}{2}B)}{\sin(\frac{1}{2}A-\frac{1}{2}B)}=\frac{\tan\frac{1}{2}[s-b+s-a]}{\tan\frac{1}{2}[s-b-(s-a)]}.$$

But
$$s-b+s-a=2s-a-b=c.$$

Whence,
$$\frac{\sin\frac{1}{2}(A+B)}{\sin\frac{1}{2}(A-B)}=\frac{\tan\frac{1}{2}c}{\tan\frac{1}{2}(a-b)}. \qquad (\mathbf{112})$$

157. Multiplying (**100**) by (**101**), we have

$$\tan\tfrac{1}{2}A\tan\tfrac{1}{2}B=\sqrt{\frac{\sin(s-b)\sin(s-c)}{\sin s\sin(s-a)}}\sqrt{\frac{\sin(s-c)\sin(s-a)}{\sin s\sin(s-b)}}.$$

Or,
$$\frac{\sin\frac{1}{2}A\sin\frac{1}{2}B}{\cos\frac{1}{2}A\cos\frac{1}{2}B}=\sqrt{\frac{\sin^2(s-c)}{\sin^2 s}}=\frac{\sin(s-c)}{\sin s}.$$

Whence by composition and division,

$$\frac{\cos\frac{1}{2}A\cos\frac{1}{2}B-\sin\frac{1}{2}A\sin\frac{1}{2}B}{\cos\frac{1}{2}A\cos\frac{1}{2}B+\sin\frac{1}{2}A\sin\frac{1}{2}B}=\frac{\sin s-\sin(s-c)}{\sin s+\sin(s-c)}.$$

Or, by (**21**),
$$\frac{\cos(\frac{1}{2}A+\frac{1}{2}B)}{\cos(\frac{1}{2}A-\frac{1}{2}B)}=\frac{\tan\frac{1}{2}[s-(s-c)]}{\tan\frac{1}{2}[s+s-c]}.$$

But
$$s+s-c=2s-c=a+b.$$

Whence,
$$\frac{\cos\frac{1}{2}(A+B)}{\cos\frac{1}{2}(A-B)}=\frac{\tan\frac{1}{2}c}{\tan\frac{1}{2}(a+b)}. \qquad (\mathbf{113})$$

158. Applying formula (**112**) to the triangle $A'B'C'$, in the figure of 153, we obtain

$$\frac{\sin\frac{1}{2}(A'+B')}{\sin\frac{1}{2}(A'-B')}=\frac{\tan\frac{1}{2}c'}{\tan\frac{1}{2}(a'-b')}.$$

But, $\frac{1}{2}(A'+B')=\frac{1}{2}(180°-a+180°-b)=180°-\frac{1}{2}(a+b)$;

$$\frac{1}{2}(A'-B')=\frac{1}{2}(180°-a-180°+b)=-\frac{1}{2}(a-b);$$

$$\frac{1}{2}c'=\frac{1}{2}(180°-C)=90°-\frac{1}{2}C;$$

and $\frac{1}{2}(a'-b')=\frac{1}{2}(180°-A-180°+B)=-\frac{1}{2}(A-B)$.

Whence, $\frac{\sin[180°-\frac{1}{2}(a+b)]}{\sin[-\frac{1}{2}(a-b)]}=\frac{\tan(90°-\frac{1}{2}C)}{\tan[-\frac{1}{2}(A-B)]}$.

Therefore, by §§ 29, 32, and 33,

$$\frac{\sin\frac{1}{2}(a+b)}{-\sin\frac{1}{2}(a-b)}=\frac{\cot\frac{1}{2}C}{-\tan\frac{1}{2}(A-B)}.$$

Or, $$\frac{\sin\frac{1}{2}(a+b)}{\sin\frac{1}{2}(a-b)}=\frac{\cot\frac{1}{2}C}{\tan\frac{1}{2}(A-B)}. \quad (114)$$

In like manner, from **(113)**, we obtain

$$\frac{\cos\frac{1}{2}(A'+B')}{\cos\frac{1}{2}(A'-B')}=\frac{\tan\frac{1}{2}c'}{\tan\frac{1}{2}(a'+b')}.$$

But, $\frac{1}{2}(a'+b')=\frac{1}{2}(180°-A+180°-B)=180°-\frac{1}{2}(A+B)$.

Whence, $\frac{\cos[180°-\frac{1}{2}(a+b)]}{\cos[-\frac{1}{2}(a-b)]}=\frac{\tan(90°-\frac{1}{2}C)}{\tan[180°-\frac{1}{2}(A+B)]}$.

Therefore, by §§ 29, 32, and 33,

$$\frac{-\cos\frac{1}{2}(a+b)}{\cos\frac{1}{2}(a-b)}=\frac{\cot\frac{1}{2}C}{-\tan\frac{1}{2}(A+B)}.$$

Or, $$\frac{\cos\frac{1}{2}(a+b)}{\cos\frac{1}{2}(a-b)}=\frac{\cot\frac{1}{2}C}{\tan\frac{1}{2}(A+B)}. \quad (115)$$

159. The formulæ exemplified in §§ 156, 157, and 158 are known as *Napier's Analogies.* In each case there may be other forms according as other elements are used.

SOLUTION OF OBLIQUE SPHERICAL TRIANGLES.

160. In the solution of oblique spherical triangles, we may distinguish six cases:

1. *Given a side and the adjacent angles.*
2. *Given two sides and their included angle.*
3. *Given the three sides.*
4. *Given the three angles.*
5. *Given two sides and the angle opposite to one of them.*
6. *Given two angles and the side opposite to one of them.*

By application of the principles of § 131, 6, the solution of an example under Case 2, 4, or 6, may be made to depend upon the solution of an example under Case 1, 3, or 5, respectively; and *vice versa.*

Hence, it is not essential to consider more than *three* cases in the solution of oblique spherical triangles.

The student must carefully bear in mind the remarks made in §§ 146 and 147.

161. Case I. *Given a side and the adjacent angles.*

1. Given $A = 70°$, $B = 132°$, $c = 116°$; find a, b, and C.

By Napier's Analogies (§§ 156, 157), we have

$$\frac{\sin\frac{1}{2}(B+A)}{\sin\frac{1}{2}(B-A)} = \frac{\tan\frac{1}{2}c}{\tan\frac{1}{2}(b-a)}, \text{ and } \frac{\cos\frac{1}{2}(B+A)}{\cos\frac{1}{2}(B-A)} = \frac{\tan\frac{1}{2}c}{\tan\frac{1}{2}(b+a)}.$$

Whence, $\tan\frac{1}{2}(b-a) = \sin\frac{1}{2}(B-A)\csc\frac{1}{2}(B+A)\tan\frac{1}{2}c,$

and $\tan\frac{1}{2}\overset{-}{(b+a)} = \cos\frac{1}{2}\overset{+}{(B-A)}\sec\frac{1}{2}\overset{-}{(B+A)}\overset{+}{\tan}\frac{1}{2}c.$

From the data, $\frac{1}{2}(B-A) = 31°$, $\frac{1}{2}(B+A) = 101°$, $\frac{1}{2}c = 58°$.

$$\begin{aligned} \log\sin\tfrac{1}{2}(B-A) &= 9.711839 - 10 \\ \log\csc\tfrac{1}{2}(B+A) &= 0.008053 \\ \log\tan\tfrac{1}{2}c &= 0.204211 \\ \hline \log\tan\tfrac{1}{2}(b-a) &= 9.924103 - 10 \\ \tfrac{1}{2}(b-a) &= 40°\,1'\,7.7''. \end{aligned}$$

$$\begin{aligned} \log\cos\tfrac{1}{2}(B-A) &= 9.933066 - 10 \\ \log\sec\tfrac{1}{2}(B+A) &= 0.719401 \\ \log\tan\tfrac{1}{2}c &= 0.204211 \\ \hline \log\tan\tfrac{1}{2}(b+a) &= 0.856678 \\ 180° - \tfrac{1}{2}(b+a) &= 82°\,4'\,51.8''. \\ \tfrac{1}{2}(b+a) &= 97°\,55'\,8.2''. \end{aligned}$$

Then, $a = \frac{1}{2}(b+a) - \frac{1}{2}(b-a) = 57°\,54'\,0.5''$,

and $b = \frac{1}{2}(b+a) + \frac{1}{2}(b-a) = 137°\,56'\,15.9''$.

To find C, we have by § 158,

$$\cot\frac{1}{2}C = \frac{\sin\frac{1}{2}(b+a)}{\sin\frac{1}{2}(b-a)}\tan\frac{1}{2}(B-A) = \sin\frac{1}{2}(b+a)\csc\frac{1}{2}(b-a)\tan\frac{1}{2}(B-A).$$

$$\begin{aligned} \log\sin\tfrac{1}{2}(b+a) &= 9.995839 - 10 \\ \log\csc\tfrac{1}{2}(b-a) &= 0.191763 \\ \log\tan\tfrac{1}{2}(B-A) &= 9.778774 - 10 \\ \hline \log\cot\tfrac{1}{2}C &= 9.966376 - 10 \\ \tfrac{1}{2}C &= 47°\,12'\,56.7'', \text{ and } C = 94°\,25'\,53.4''. \end{aligned}$$

Note 1. The value of C may also be found by the formula

$$\cot\frac{1}{2}C = \frac{\cos\frac{1}{2}(b+a)}{\cos\frac{1}{2}(b-a)}\tan\frac{1}{2}(B+A) \quad (\S\ 158).$$

Note 2. The triangle is possible for any values of the given elements.

EXAMPLES.

Solve the following spherical triangles:

2. Given $A = 78°$, $B = 41°$, $c = 108°$.

3. Given $B = 115°$, $C = 50°$, $a = 70° 20'$.

4. Given $A = 31° 40'$, $C = 122° 20'$, $b = 40° 40'$.

5. Given $A = 108° 12'$, $B = 145° 46'$, $c = 126° 32'$.

162. Case II. *Given two sides and their included angle.*

1. Given $b = 138°$, $c = 116°$, $A = 70°$; find B, C, and a.

By Napier's Analogies (§ 158), we have

$$\frac{\sin \frac{1}{2}(b+c)}{\sin \frac{1}{2}(b-c)} = \frac{\cot \frac{1}{2}A}{\tan \frac{1}{2}(B-C)}, \text{ and } \frac{\cos \frac{1}{2}(b+c)}{\cos \frac{1}{2}(b-c)} = \frac{\cot \frac{1}{2}A}{\tan \frac{1}{2}(B+C)}.$$

Whence, $\tan \frac{1}{2}(B-C) = \sin \frac{1}{2}(b-c) \csc \frac{1}{2}(b+c) \cot \frac{1}{2}A$,

and $$\tan \frac{1}{2}\overset{-}{(B+C)} = \cos \frac{1}{2}\overset{+}{(b-c)} \sec \frac{1}{2}\overset{-}{(b+c)} \cot \overset{+}{\frac{1}{2}A}.$$

From the data, $\frac{1}{2}(b-c) = 11°$, $\frac{1}{2}(b+c) = 127°$, $\frac{1}{2}A = 35°$.

$$\begin{aligned} \log \sin \tfrac{1}{2}(b-c) &= 9.280599 - 10 \\ \log \csc \tfrac{1}{2}(b+c) &= 0.097651 \\ \log \cot \tfrac{1}{2}A &= 0.154773 \\ \hline \log \tan \tfrac{1}{2}(B-C) &= 9.533023 - 10 \\ \tfrac{1}{2}(B-C) &= 18° 50' 24.7''. \end{aligned}$$

$$\begin{aligned} \log \cos \tfrac{1}{2}(b-c) &= 9.991947 - 10 \\ \log \sec \tfrac{1}{2}(b+c) &= 0.220537 \\ \log \cot \tfrac{1}{2}A &= 0.154773 \\ \hline \log \tan \tfrac{1}{2}(B+C) &= 0.367257 \\ 180° - \tfrac{1}{2}(B+C) &= 66° 46' 1.2''. \\ \tfrac{1}{2}(B+C) &= 113° 13' 58.8''. \end{aligned}$$

Then, $$B = \tfrac{1}{2}(B+C) + \tfrac{1}{2}(B-C) = 132° 4' 23.5'',$$

and $$C = \tfrac{1}{2}(B+C) - \tfrac{1}{2}(B-C) = 94° 23' 34.1''.$$

To find a, we have by § 156,

$$\tan \tfrac{1}{2}a = \frac{\sin \frac{1}{2}(B+C)}{\sin \frac{1}{2}(B-C)} \tan \tfrac{1}{2}(b-c) = \sin \tfrac{1}{2}(B+C) \csc \tfrac{1}{2}(B-C) \tan \tfrac{1}{2}(b-c).$$

$$\begin{aligned} \log \sin \tfrac{1}{2}(B+C) &= 9.963272 - 10 \\ \log \csc \tfrac{1}{2}(B-C) &= 0.490892 \\ \log \tan \tfrac{1}{2}(b-c) &= 9.288652 - 10 \\ \hline \log \tan \tfrac{1}{2}a &= 9.742816 - 10 \end{aligned}$$

$\frac{1}{2}a = 28° 56' 51.6''$, and $a = 57° 53' 43.2''$.

Note. The triangle is possible for any values of the given elements.

EXAMPLES.

Solve the following spherical triangles:

2. Given $a = 72°$, $b = 47°$, $C = 33°$.
3. Given $a = 98°$, $c = 60°$, $B = 110°$.
4. Given $b = 70° 40'$, $c = 120° 20'$, $A = 50°$.
5. Given $a = 125° 10'$, $b = 153° 50'$, $C = 140° 20'$.

163. Case III. *Given the three sides.*

The angles may be calculated by the formulæ of § 154.

If all the angles are to be computed, the *tangent* formulæ are the most convenient, since only four different logarithms are required. If but one angle is required, the *cosine* formula will be found to involve the least work.

The triangle is possible for any values of the data, provided that no side is greater than the sum of the other two, and that the sum of the sides is less than 360° (§ 131, 1 and 3).

If all the angles are required, and the tangent formulæ are used, it is convenient to modify them as follows.

By (100), $$\tan \tfrac{1}{2} A = \sqrt{\frac{\sin(s-a)\sin(s-b)\sin(s-c)}{\sin s \sin^2(s-a)}}$$

$$= \frac{1}{\sin(s-a)}\sqrt{\frac{\sin(s-a)\sin(s-b)\sin(s-c)}{\sin s}}.$$

Denoting $\sqrt{\frac{\sin(s-a)\sin(s-b)\sin(s-c)}{\sin s}}$ by k, we have

$$\tan \tfrac{1}{2} A = \frac{k}{\sin(s-a)}.$$

In like manner, $\tan \tfrac{1}{2} B = \frac{k}{\sin(s-b)}$, and $\tan \tfrac{1}{2} C = \frac{k}{\sin(s-c)}$.

1. Given $a = 57°$, $b = 137°$, $c = 116°$; find A, B, and C.

Here, $$2s = a + b + c = 310°.$$

Whence, $s = 155°$, $s - a = 98°$, $s - b = 18°$, $s - c = 39°$.

$$\begin{aligned}
\log \sin(s-a) &= 9.995753 - 10\\
\log \sin(s-b) &= 9.489982 - 10\\
\log \sin(s-c) &= 9.798872 - 10\\
\log \csc s &= 0.374052\\
2)&\underline{19.658659 - 20}\\
\log k &= 9.829330 - 10\\
\log \sin(s-a) &= 9.995753 - 10\\
\log \tan \tfrac{1}{2} A &= 9.833577 - 10\\
\tfrac{1}{2} A &= 34° 16' 52.5''.\\
A &= 68° 33' 45.0''.
\end{aligned}$$

$$\begin{aligned}
\log k &= 9.829330 - 10\\
\log \sin(s-b) &= 9.489982 - 10\\
\log \tan \tfrac{1}{2} B &= 0.339348\\
\tfrac{1}{2} B &= 65° 24' 10.4''.\\
B &= 130° 48' 20.8''\\
\log k &= 9.829330 - 10\\
\log \sin(s-c) &= 9.798872 - 10\\
\log \tan \tfrac{1}{2} C &= 0.030458\\
\tfrac{1}{2} C &= 47° 0' 27.0''.\\
C &= 94° 0' 54.0''.
\end{aligned}$$

EXAMPLES.

Solve the following spherical triangles:

2. Given $a = 38°$, $b = 42°$, $c = 51°$.

3. Given $a = 101°$, $b = 49°$, $c = 60°$.

4. Given $a = 126°$, $b = 152°$, $c = 75°$.

5. Given $a = 62° 20'$, $b = 54° 10'$, $c = 97° 50'$; find A.

164. CASE IV. *Given the three angles.*

The sides may be calculated by the formulæ of § 155.

If all the sides are to be computed, the tangent formulæ are the most convenient, since only four different logarithms are required. If but one side is required, the sine formula will be found to involve the least work.

The triangle is possible for any values of the data, provided that the sum of the angles is between 180° and 540° (§ 131, 4), and that each of the quantities $B + C - A$, $C + A - B$, and $A + B - C$ is between 180° and $-180°$ (§ 134).

For such values of the angles, S is between 90° and 270°, and each of the quantities $S - A$, $S - B$, and $S - C$ between 90° and $-90°$; then, $\cos S$ is $-$, while the cosines of $S - A$, $S - B$, and $S - C$ are $+$ (§ 20).

Hence, the expressions under the radical signs in the formulæ are essentially positive, and no attention need be paid to the algebraic signs.

If all the sides are required, and the tangent formulæ are used, it is convenient to modify them as follows:

By (**109**),
$$\tan \tfrac{1}{2} a = \sqrt{-\frac{\cos S \cos^2 (S - A)}{\cos (S - A) \cos (S - B) \cos (S - C)}}$$

$$= \cos (S - A) \sqrt{-\frac{\cos S}{\cos (S - A) \cos (S - B) \cos (S - C)}}.$$

Denoting $\sqrt{-\frac{\cos S}{\cos (S - A) \cos (S - B) \cos (S - C)}}$ by K,

we have
$$\tan \tfrac{1}{2} a = K \cos (S - A).$$

In like manner, $\tan \tfrac{1}{2} b = K \cos (S - B)$, and $\tan \tfrac{1}{2} c = K \cos (S - C)$.

1. Given $A = 150°$, $B = 131°$, $C = 115°$; find a, b, and c.

Here,
$$2S = A + B + C = 396°.$$

Whence, $S = 198°$, $S - A = 48°$, $S - B = 67°$, $S - C = 83°$.

$$\begin{aligned}\log\cos S &= 9.978206-10\\ \log\sec(S-A) &= 0.174489\\ \log\sec(S-B) &= 0.408122\\ \log\sec(S-C) &= 0.914106\\ &2\overline{)1.474923}\\ \log K &= 0.737462\\ \log\cos(S-A) &= 9.825511-10\\ \log\tan\tfrac{1}{2}a &= 0.562973\\ \tfrac{1}{2}a &= 74^\circ\,42'\,4.8''.\\ a &= 149^\circ\,24'\,9.6''.\end{aligned}$$

$$\begin{aligned}\log K &= 0.737462\\ \log\cos(S-B) &= 9.591878-10\\ \log\tan\tfrac{1}{2}b &= 0.329340\\ \tfrac{1}{2}b &= 64^\circ\,53'\,58.0''.\\ b &= 129^\circ\,47'\,56.0''.\\ \log K &= 0.737462\\ \log\cos(S-C) &= 9.085894-10\\ \log\tan\tfrac{1}{2}c &= 9.823356-10\\ \tfrac{1}{2}c &= 33^\circ\,39'\,23.1''.\\ c &= 67^\circ\,18'\,46.2''.\end{aligned}$$

Note 1. By § 35, $\cos 198^\circ = -\sin 108^\circ = -\cos 18^\circ$; whence, without regard to algebraic sign, $\log\cos 198^\circ = \log\cos 18^\circ$.

2. Given $A = 123^\circ$, $B = 45^\circ$, $C = 58^\circ$; find a.

By (103), $$\sin\tfrac{1}{2}a = \sqrt{-\frac{\cos S\cos(S-A)}{\sin B\sin C}}.$$

Here, $2S = A + B + C = 226^\circ$; whence, $S = 113^\circ$, and $S - A = -10^\circ$.

$$\begin{aligned}\log\cos S &= 9.591878-10\\ \log\cos(S-A) &= 9.993351-10\\ \log\csc B &= 0.150515\\ \log\csc C &= 0.071580\\ &2\overline{)19.807324-20}\\ \log\sin\tfrac{1}{2}a &= 9.903662-10\end{aligned}$$

$$\tfrac{1}{2}a = 53^\circ\,13'\,51.3'', \text{ and } a = 106^\circ\,27'\,42.6''.$$

Note 2. By § 29, $\cos(-10^\circ) = \cos 10^\circ$.

EXAMPLES.

Solve the following spherical triangles:

3. Given $A = 74^\circ$, $B = 82^\circ$, $C = 67^\circ$.
4. Given $A = 120^\circ$, $B = 130^\circ$, $C = 140^\circ$.
5. Given $A = 138^\circ\,16'$, $B = 33^\circ\,11'$, $C = 36^\circ\,53'$.
6. Given $A = 91^\circ\,10'$, $B = 85^\circ\,40'$, $C = 78^\circ\,30'$; find b.

165. CASE V. *Given two sides and the angle opposite to one of them.*

1. Given $a = 58^\circ$, $b = 137^\circ$, $B = 131^\circ$; find A, C, and c.

By (85), $$\frac{\sin A}{\sin B} = \frac{\sin a}{\sin b}, \text{ or } \sin A = \sin a \csc b \sin B.$$

$$\log \sin a = 9.928420 - 10$$
$$\log \csc b = 0.166217$$
$$\log \sin B = 9.877780 - 10$$
$$\log \sin A = 9.972417 - 10$$
$$A = 69^\circ 47' 41.6'', \text{ or } 110^\circ 12' 18.4'' \text{ (§ 147)}.$$

To find C and c, we have by §§ 156 and 158,

$$\cot \tfrac{1}{2} C = \sin \tfrac{1}{2}(b+a) \csc \tfrac{1}{2}(b-a) \tan \tfrac{1}{2}(B-A),$$

and $$\tan \tfrac{1}{2} c = \sin \tfrac{1}{2}(B+A) \csc \tfrac{1}{2}(B-A) \tan \tfrac{1}{2}(b-a).$$

Using the first value of A, we have

$$\tfrac{1}{2}(B+A) = 100^\circ 23' 50.8'', \text{ and } \tfrac{1}{2}(B-A) = 30^\circ 36' 9.2''.$$

Also, $\tfrac{1}{2}(b+a) = 97^\circ 30'$, and $\tfrac{1}{2}(b-a) = 39^\circ 30'$.

$$\log \sin \tfrac{1}{2}(b+a) = 9.996269 - 10$$
$$\log \csc \tfrac{1}{2}(b-a) = 0.196489$$
$$\log \tan \tfrac{1}{2}(B-A) = 9.771924 - 10$$
$$\log \cot \tfrac{1}{2} C = 9.964682 - 10$$
$$\tfrac{1}{2} C = 47^\circ 19' 37.8''.$$
$$C = 94^\circ 39' 15.6''.$$

$$\log \sin \tfrac{1}{2}(B+A) = 9.992810 - 10$$
$$\log \csc \tfrac{1}{2}(B-A) = 0.293214$$
$$\log \tan \tfrac{1}{2}(b-a) = 9.916104 - 10$$
$$\log \tan \tfrac{1}{2} c = 0.202128$$
$$\tfrac{1}{2} c = 57^\circ 52' 35.0''.$$
$$c = 115^\circ 45' 10.0''.$$

Using the second value of A, we have

$$\tfrac{1}{2}(B+A) = 120^\circ 36' 9.2'', \text{ and } \tfrac{1}{2}(B-A) = 10^\circ 23' 50.8''.$$

$$\log \sin \tfrac{1}{2}(b+a) = 9.996269 - 10$$
$$\log \csc \tfrac{1}{2}(b-a) = 0.196489$$
$$\log \tan \tfrac{1}{2}(B-A) = 9.263608 - 10$$
$$\log \cot \tfrac{1}{2} C = 9.456366 - 10$$
$$\tfrac{1}{2} C = 74^\circ 2' 22.1''.$$
$$C = 148^\circ 4' 44.2''.$$

$$\log \sin \tfrac{1}{2}(B+A) = 9.934861 - 10$$
$$\log \csc \tfrac{1}{2}(B-A) = 0.743583$$
$$\log \tan \tfrac{1}{2}(b-a) = 9.916104 - 10$$
$$\log \tan \tfrac{1}{2} c = 0.594548$$
$$\tfrac{1}{2} c = 75^\circ 43' 43.6''.$$
$$c = 151^\circ 27' 27.2''.$$

Thus, the two solutions are:

1. $A = 69^\circ 47' 41.6''$, $C = 94^\circ 39' 15.6''$, $c = 115^\circ 45' 10.0''$.
2. $A = 110^\circ 12' 18.4''$, $C = 148^\circ\ 4' 44.2''$, $c = 151^\circ 27' 27.2''$.

As in the corresponding case in the solution of oblique plane triangles (compare §§ 117 to 120), there may sometimes be two solutions, sometimes only one, and sometimes none, in an example under Case V.

After the two values of A have been obtained, the number of solutions may be readily determined by inspection; for, by § 131, 2, if a is $< b$, A must be $< B$; and if a is $> b$, A must be $> B$.

Hence, *only those values of A can be retained which are greater or less than B according as a is greater or less than b.*

Thus, in Ex. 1, a is given $< b$; and since both values of A are $< B$, we have two solutions.

Again, if the data are such as to make log sin A positive, there will be no solution corresponding.

2. Given $a = 58°$, $c = 116°$, $C = 94° 50'$; find A.

In this case, $\frac{\sin A}{\sin C} = \frac{\sin a}{\sin c}$, or $\sin A = \sin a \csc c \sin C$.

$$\begin{aligned} \log \sin a &= 9.928420 - 10 \\ \log \csc c &= 0.046340 \\ \log \sin C &= \underline{9.998453 - 10} \\ \log \sin A &= 9.973213 - 10 \end{aligned}$$

$$A = 70° 4' 57.1'', \text{ or } 109° 55' 2.9''.$$

Since a is given $< c$, only values of A which are $< C$ can be retained; then the only solution is $A = 70° 4' 57.1''$.

3. Given $b = 126°$, $c = 70°$, $B = 56°$; find C.

In this case, $\frac{\sin C}{\sin B} = \frac{\sin c}{\sin b}$, or $\sin C = \sin c \csc b \sin B$.

$$\begin{aligned} \log \sin c &= 9.972986 - 10 \\ \log \csc b &= 0.092042 \\ \log \sin B &= \underline{9.918574 - 10} \\ \log \sin C &= 9.983602 - 10 \end{aligned}$$

$$C = 74° 21' 13.8'', \text{ or } 105° 38' 46.2''.$$

Since both values of C are $> B$, while c is given $< b$, there is no solution.

EXAMPLES.

Solve the following spherical triangles:

4. Given $b = 99° 40'$, $c = 64° 20'$, $B = 95° 40'$.

5. Given $a = 40°$, $b = 118° 20'$, $A = 29° 40'$.

6. Given $a = 115° 20'$, $c = 146° 20'$, $C = 141° 10'$.

7. Given $a = 109° 20'$, $c = 82° 1' 8''$, $A = 107° 40'$.

8. Given $b = 108° 30'$, $c = 40° 50'$, $C = 39° 50'$.

9. Given $a = 162° 20'$, $b = 15° 40'$, $B = 125°$.

10. Given $a = 55°$, $c = 138° 10'$, $A = 42° 30'$.

166. CASE VI. *Given two angles and the side opposite to one of them*

1. Given $A=110°$, $B=122°$, $b=129°$; find a, c, and C.

By (**85**), $\frac{\sin a}{\sin b}=\frac{\sin A}{\sin B}$, or $\sin a = \sin A \csc B \sin b$.

$$\begin{aligned}\log\sin A &= 9.972986-10\\ \log\csc B &= 0.071580\\ \log\sin b &= 9.890503-10\\ \hline \log\sin a &= 9.935069-10\end{aligned}$$

$$a = 59°\,26'\,37.6'', \text{ or } 120°\,33'\,22.4''\ (\S\,147).$$

To find c and C, we have by §§ 156 and 158,

$$\tan\tfrac{1}{2}c = \sin\tfrac{1}{2}(B+A)\csc\tfrac{1}{2}(B-A)\tan\tfrac{1}{2}(b-a),$$

and $$\cot\tfrac{1}{2}C = \sin\tfrac{1}{2}(b+a)\csc\tfrac{1}{2}(b-a)\tan\tfrac{1}{2}(B-A).$$

Using the first value of a, we have

$$\tfrac{1}{2}(b+a)=94°\,13'\,18.8'', \text{ and } \tfrac{1}{2}(b-a)=34°\,46'\,41.2''.$$

Also, $$\tfrac{1}{2}(B+A)=116°, \text{ and } \tfrac{1}{2}(B-A)=6°.$$

$$\begin{aligned}\log\sin\tfrac{1}{2}(B+A) &= 9.953660-10\\ \log\csc\tfrac{1}{2}(B-A) &= 0.980765\\ \log\tan\tfrac{1}{2}(b-a) &= 9.841642-10\\ \hline \log\tan\tfrac{1}{2}c &= 0.776067\\ \tfrac{1}{2}c &= 80°\,29'\,34.8''.\\ c &= 160°\,59'\,9.6''.\end{aligned}$$

$$\begin{aligned}\log\sin\tfrac{1}{2}(b+a) &= 9.998820-10\\ \log\csc\tfrac{1}{2}(b-a) &= 0.243821\\ \log\tan\tfrac{1}{2}(B-A) &= 9.021620-10\\ \hline \log\cot\tfrac{1}{2}C &= 9.264261-10\\ \tfrac{1}{2}C &= 79°\,35'\,14.1''.\\ C &= 159°\,10'\,28.2''.\end{aligned}$$

Using the second value of a, we have

$$\tfrac{1}{2}(b+a)=124°\,46'\,41.2'', \text{ and } \tfrac{1}{2}(b-a)=4°\,13'\,18.8''.$$

$$\begin{aligned}\log\sin\tfrac{1}{2}(B+A) &= 9.953660-10\\ \log\csc\tfrac{1}{2}(B-A) &= 0.980765\\ \log\tan\tfrac{1}{2}(b-a) &= 8.868171-10\\ \hline \log\tan\tfrac{1}{2}c &= 9.802596-10\\ \tfrac{1}{2}c &= 32°\,24'\,17.8''.\\ c &= 64°\,48'\,35.6''.\end{aligned}$$

$$\begin{aligned}\log\sin\tfrac{1}{2}(b+a) &= 9.914537-10\\ \log\csc\tfrac{1}{2}(b-a) &= 1.133009\\ \log\tan\tfrac{1}{2}(B-A) &= 9.021620-10\\ \hline \log\cot\tfrac{1}{2}C &= 0.069166\\ \tfrac{1}{2}C &= 40°\,27'\,24.1''.\\ C &= 80°\,54'\,48.2''.\end{aligned}$$

Thus the two solutions are:

1. $a = 59°\,26'\,37.6''$, $c=160°\,59'\,9.6''$, $C=159°\,10'\,28.2''$.
2. $a=120°\,33'\,22.4''$, $c=64°\,48'\,35.6''$, $C=80°\,54'\,48.2''$.

In examples in Case VI., as well as in Case V., there may sometimes be two solutions, sometimes only one, and sometimes none.

As in Case V., *only those values of a can be retained which are greater or less than b according as A is greater or less than B.*

Also, if log sin a is positive, the triangle is impossible.

EXAMPLES.

Solve the following spherical triangles:

2. Given $B = 116°$, $C = 80°$, $c = 84°$.

3. Given $A = 132°$, $B = 140°$, $b = 127°$.

4. Given $A = 62°$, $C = 101° 58' 24''$, $a = 64° 30'$.

5. Given $A = 133° 50'$, $B = 66° 30'$, $a = 81° 10'$.

6. Given $B = 22° 20'$, $C = 146° 40'$, $c = 138° 20'$.

7. Given $A = 61° 40'$, $C = 140° 20'$, $c = 150° 20'$.

8. Given $B = 73°$, $C = 81° 20'$, $b = 122° 40'$.

APPLICATIONS.

167. In problems concerning navigation, the earth may be regarded as a sphere.

The *shortest distance* between any two points on the surface is the arc of a great circle which joins them; and the angles between this arc and the meridians of the points determine the *bearings* of the points from each other.

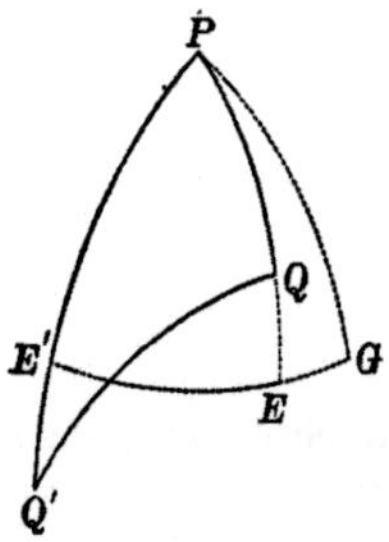

Thus, if Q and Q' are the points, and PQ and PQ' their meridians, the angle PQQ' determines the bearing of Q' from Q, and the angle $PQ'Q$ determines the bearing of Q from Q'.

If the latitudes and longitudes of Q and Q' are known, the arc QQ' and the angles PQQ' and $PQ'Q$ may be determined by the solution of a spherical triangle.

For if EE' is the equator, and PG the meridian of Greenwich, we have

$$\angle QPQ' = \angle Q'PG - \angle QPG = \text{longitude } Q' - \text{longitude } Q.$$

Also, $PQ = PE - QE = 90° - \text{latitude } Q,$

and $PQ' = PE' + Q'E' = 90° + \text{latitude } Q'.$

Thus, in the spherical triangle PQQ', two sides and their included angle are known, and the remaining elements may be computed.

When QQ' has been found in degrees, its length in miles may be calculated by finding the ratio of its arc to 360°, and multiplying the result by the length of the circumference of a great circle; in the following problems, the radius of the earth is taken as 3956 miles.

EXAMPLES.

1. Boston lies in lat. 42° 21′ N., lon. 71° 4′ W.; and the latitude of Greenwich is 51° 29′ N. Find the shortest distance in miles between the places, and the bearing of each place from the other.

2. Calcutta lies in lat. 22° 33′ N., lon. 88° 19′ E.; and Valparaiso in lat. 33° 2′ S., lon. 71° 42′ W. Find the shortest distance in miles between the places, and the bearing of each place from the other.

3. Sandy Hook lies in lat. 40° 28′ N., lon. 74° 1′ W.; and Queenstown in lat. 51° 50′ N., lon. 8° 19′ W. In what latitude does a great circle course from Sandy Hook to Queenstown cross the meridian of 50° W.?

168. The Astronomical Triangle.

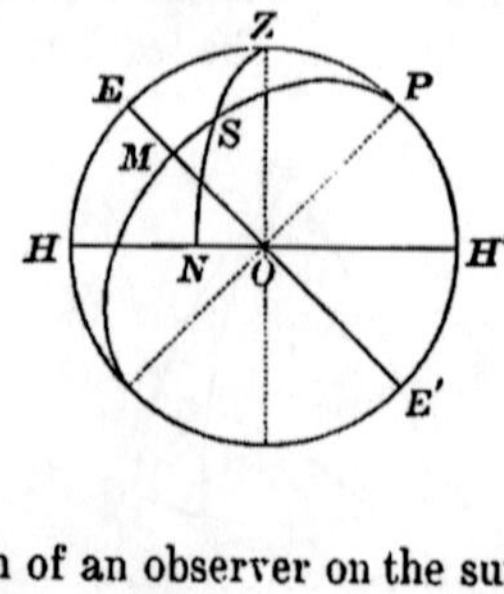

Let O be the position of an observer on the surface of the earth; P the celestial north-pole; Z the zenith.

The great circle EE', having P for its pole, is called the *celestial equator;* and the great circle HH', having Z for its pole, is called the *horizon.*

Let S be the position of a star; PSM a meridian passing through S; ZSN a quadrant of a great circle passing through Z and S.

The arc SM is called the *declination* of the star; and is called declination *north* or *south*, according as the star is north or south of the celestial equator.

The angle SPZ is called the *hour-angle* of the star; the arc SN is called its *altitude;* the angle PZS, its *bearing* or *azimuth.*

The arc EZ is the latitude of the observer.

The spherical triangle SPZ is called the *Astronomical Triangle;* its sides have the following values:

$$SP = PM - SM = 90° - \text{the declination of the star;}$$

$$SZ = ZN - SN = 90° - \text{the altitude of the star;}$$

$$PZ = EP - EZ = 90° - \text{the latitude of the observer.}$$

Its angle SPZ is the hour-angle of the star, and its angle SZP the azimuth.

If any three of these five elements are known, the solution of a spherical triangle serves to determine the other two.

169. To Determine the Hour of the Day.

If the altitude and declination of the sun are known, and the latitude of the observer, the three sides of the triangle SPZ are known, and the hour-angle SPZ may be computed.

If 24 hours be multiplied by the ratio of this angle to 360°, we obtain the time required for the sun to move from S to the meridian EP.

If this time be subtracted from 12 o'clock, if the observation is made in the morning, or added, if made in the afternoon, we obtain the *hour of the day* at the time and place of observation.

If the Greenwich time of the observation be noted on a chronometer, the difference between this and the local time as calculated above serves to determine the *longitude* of the place of observation.

In reducing time to longitude, it should be remembered that 24 hours of time correspond to 360° of longitude; that is, one hour of time corresponds to 15° of longitude, one minute to 15′, and one second to 15″.

EXAMPLES.

170. 1. A mariner observes the altitude of the sun to be 14° 18′, its declination being 18° 36′ N. If the latitude of the vessel is 50° 13′ N., and the observation is made in the morning, find the hour of the day. If the observation is taken at 9 A.M., Greenwich time, what is the longitude of the vessel?

2. What will be the altitude of the sun at 4 P.M. in San Francisco, lat. 37° 48′ N., its declination being 12° S.?

3. What will be the bearing of the sun at 9.30 A.M. in Melbourne, lat. 37° 49′ S., if its declination is 6° S.?

4. At what hour will the sun rise in Boston, lat. 42° 21′ N., if its declination is 15° N.?

Note. At sunrise the sun's altitude is 0, so that the arc SZ becomes 90°.

FORMULÆ.

PLANE TRIGONOMETRY.

§ 29. $\sin(-A) = -\sin A. \quad \tan(-A) = -\tan A. \quad \sec(-A) = \sec A.$
$\cos(-A) = \cos A. \quad \cot(-A) = -\cot A. \quad \csc(-A) = -\csc A.$ (1)

§ 30. $\sin(90° + A) = \cos A. \quad \cot(90° + A) = -\tan A.$
$\cos(90° + A) = -\sin A. \quad \sec(90° + A) = -\csc A.$
$\tan(90° + A) = -\cot A. \quad \csc(90° + A) = \sec A.$ (2)

§ 36. $\sin x = \frac{1}{\csc x}. \quad \tan x = \frac{1}{\cot x}. \quad \sec x = \frac{1}{\cos x}.$
$\cos x = \frac{1}{\sec x}. \quad \cot x = \frac{1}{\tan x}. \quad \csc x = \frac{1}{\sin x}.$ (3)

§ 37. $\tan x = \frac{\sin x}{\cos x}.$ (4)

§ 38. $\cot x = \frac{\cos x}{\sin x}.$ (5)

§ 39. $\sin^2 x + \cos^2 x = 1.$ (6)

§ 40. $\sec^2 x = 1 + \tan^2 x.$ (7) $\csc^2 x = 1 + \cot^2 x.$ (8)

§ 41 $\sin(x + y) = \sin x \cos y + \cos x \sin y.$ (9)

$\cos(x + y) = \cos x \cos y - \sin x \sin y.$ (10)

§ 43. $\sin(x - y) = \sin x \cos y - \cos x \sin y.$ (11)

$\cos(x - y) = \cos x \cos y + \sin x \sin y.$ (12)

§ 44. $\tan(x+y) = \frac{\tan x + \tan y}{1 - \tan x \tan y}.$ (13) $\cot(x+y) = \frac{\cot x \cot y - 1}{\cot y + \cot x}.$ (15)

$\tan(x-y) = \frac{\tan x - \tan y}{1 + \tan x \tan y}.$ (14) $\cot(x-y) = \frac{\cot x \cot y + 1}{\cot y - \cot x}.$ (16)

§ 45. $\sin x + \sin y = 2 \sin \tfrac{1}{2}(x + y) \cos \tfrac{1}{2}(x - y).$ (17)

$\sin x - \sin y = 2 \cos \tfrac{1}{2}(x + y) \sin \tfrac{1}{2}(x - y).$ (18)

$\cos x + \cos y = 2 \cos \tfrac{1}{2}(x + y) \cos \tfrac{1}{2}(x - y).$ (19)

$\cos x - \cos y = -2 \sin \tfrac{1}{2}(x + y) \sin \tfrac{1}{2}(x - y).$ (20)

§ 46. $\frac{\sin x + \sin y}{\sin x - \sin y} = \frac{\tan \frac{1}{2}(x + y)}{\tan \frac{1}{2}(x - y)}.$ (21)

§ 47. $\sin(x + y) \sin(x - y) = \sin^2 x - \sin^2 y.$ (22)

$\sin(x + y) \sin(x - y) = \cos^2 y - \cos^2 x.$ (23)

$\cos(x + y) \cos(x - y) = \cos^2 x - \sin^2 y = \cos^2 y - \sin^2 x.$ (24)

§ 48. $\sin 2x = 2 \sin x \cos x.$ (25) $\cos 2x = 2\cos^2 x - 1.$ (28)

$\cos 2x = \cos^2 x - \sin^2 x.$ (26) $\tan 2x = \frac{2 \tan x}{1 - \tan^2 x}.$ (29)

$\cos 2x = 1 - 2 \sin^2 x.$ (27) $\cot 2x = \frac{\cot^2 x - 1}{2 \cot x}.$ (30)

§ 49. $2 \sin^2 \frac{1}{2} x = 1 - \cos x.$ (31) $\tan \frac{1}{2} x = \frac{1 - \cos x}{\sin x}.$ (33)

$2 \cos^2 \frac{1}{2} x = 1 + \cos x.$ (32) $\cot \frac{1}{2} x = \frac{1 + \cos x}{\sin x}.$ (34)

§ 50. $\sin 3x = 3 \sin x - 4 \sin^3 x.$ (35) $\cos 3x = 4 \cos^3 x - 3 \cos x.$ (36)

$$\tan 3x = \frac{3 \tan x - \tan^3 x}{1 - 3 \tan^2 x}. \quad (37)$$

§ 105. $4K = c^2 \sin 2A.$ (38) $2K = a^2 \tan B.$ (42)

$4K = c^2 \sin 2B.$ (39) $2K = b^2 \tan A.$ (43)

$2K = a^2 \cot A.$ (40) $2K = a\sqrt{(c+a)(c-a)}.$ (44)

$2K = b^2 \cot B.$ (41) $2K = b\sqrt{(c+b)(c-b)}.$ (45)

$$2K = ab. \quad (46)$$

§ 107.

$$a : b = \sin A : \sin B. \quad (47)$$

$$b : c = \sin B : \sin C. \quad (48)$$

$$c : a = \sin C : \sin A. \quad (49)$$

§ 108.

$$\frac{a+b}{a-b} = \frac{\tan \frac{1}{2}(A+B)}{\tan \frac{1}{2}(A-B)}. \quad (50)$$

$$\frac{b+c}{b-c} = \frac{\tan \frac{1}{2}(B+C)}{\tan \frac{1}{2}(B-C)}. \quad (51)$$

$$\frac{c+a}{c-a} = \frac{\tan \frac{1}{2}(C+A)}{\tan \frac{1}{2}(C-A)}. \quad (52)$$

§ 109.

$$a^2 = b^2 + c^2 - 2bc \cos A. \quad (53)$$

$$b^2 = c^2 + a^2 - 2ca \cos B. \quad (54)$$

$$c^2 = a^2 + b^2 - 2ab \cos C. \quad (55)$$

§ 110. $\cos A = \frac{b^2 + c^2 - a^2}{2bc}.$ (56) $\cos B = \frac{c^2 + a^2 - b^2}{2ca}.$ (57)

$$\cos C = \frac{a^2 + b^2 - c^2}{2ab}. \quad (58)$$

§ 111.

$$\sin \frac{1}{2} A = \sqrt{\frac{(s-b)(s-c)}{bc}}. \quad (59)$$

$$\sin \frac{1}{2} B = \sqrt{\frac{(s-c)(s-a)}{ca}}. \quad (60)$$

$$\sin \tfrac{1}{2} C = \sqrt{\frac{(s-a)(s-b)}{ab}}. \tag{61}$$

$$\cos \tfrac{1}{2} A = \sqrt{\frac{s(s-a)}{bc}}. \tag{62}$$

$$\cos \tfrac{1}{2} B = \sqrt{\frac{s(s-b)}{ca}}. \tag{63}$$

$$\cos \tfrac{1}{2} C = \sqrt{\frac{s(s-c)}{ab}}. \tag{64}$$

$$\tan \tfrac{1}{2} A = \sqrt{\frac{(s-b)(s-c)}{s(s-a)}}. \tag{65}$$

$$\tan \tfrac{1}{2} B = \sqrt{\frac{(s-c)(s-a)}{s(s-b)}}. \tag{66}$$

$$\tan \tfrac{1}{2} C = \sqrt{\frac{(s-a)(s-b)}{s(s-c)}}. \tag{67}$$

§ 112. $2K = bc \sin A.$ (68) $\quad 2K = \dfrac{a^2 \sin B \sin C}{\sin A}.$ (71)

$2K = ca \sin B.$ (69) $\quad 2K = \dfrac{b^2 \sin C \sin A}{\sin B}.$ (72)

$2K = ab \sin C.$ (70) $\quad 2K = \dfrac{c^2 \sin A \sin B}{\sin C}.$ (73)

$$K = \sqrt{s(s-a)(s-b)(s-c)}. \tag{74}$$

SPHERICAL TRIGONOMETRY.

§ 135. $\cos c = \cos a \cos b.$ (75)

$\sin A = \dfrac{\sin a}{\sin c}.$ (76) $\quad \sin B = \dfrac{\sin b}{\sin c}.$ (78)

$\cos A = \dfrac{\tan b}{\tan c}.$ (77) $\quad \cos B = \dfrac{\tan a}{\tan c}.$ (79)

§ 137. $\tan A = \dfrac{\tan a}{\sin b}.$ (80) $\quad \tan B = \dfrac{\tan b}{\sin a}.$ (81)

§ 138. $\sin A = \dfrac{\cos B}{\cos b}.$ (82) $\quad \sin B = \dfrac{\cos A}{\cos a}.$ (83)

§ 139. $\cos c = \cot A \cot B.$ (84)

§ 151. $\dfrac{\sin A}{\sin B} = \dfrac{\sin a}{\sin b}.$ (85)

$$\frac{\sin B}{\sin C} = \frac{\sin b}{\sin c}. \tag{86}$$

$$\frac{\sin A}{\sin C} = \frac{\sin a}{\sin c}. \tag{87}$$

§ 152. $$\cos a = \cos b \cos c + \sin b \sin c \cos A. \tag{88}$$

$$\cos b = \cos c \cos a + \sin c \sin a \cos B. \tag{89}$$

$$\cos c = \cos a \cos b + \sin a \sin b \cos C. \tag{90}$$

§ 153. $$\cos A = -\cos B \cos C + \sin B \sin C \cos a. \tag{91}$$

$$\cos B = -\cos C \cos A + \sin C \sin A \cos b. \tag{92}$$

$$\cos C = -\cos A \cos B + \sin A \sin B \cos c. \tag{93}$$

§ 154. $$\sin \tfrac{1}{2} A = \sqrt{\frac{\sin (s-b) \sin (s-c)}{\sin b \sin c}}. \tag{94}$$

$$\sin \tfrac{1}{2} B = \sqrt{\frac{\sin (s-c) \sin (s-a)}{\sin c \sin a}}. \tag{95}$$

$$\sin \tfrac{1}{2} C = \sqrt{\frac{\sin (s-a) \sin (s-b)}{\sin a \sin b}}. \tag{96}$$

$$\cos \tfrac{1}{2} A = \sqrt{\frac{\sin s \sin (s-a)}{\sin b \sin c}}. \tag{97}$$

$$\cos \tfrac{1}{2} B = \sqrt{\frac{\sin s \sin (s-b)}{\sin c \sin a}}. \tag{98}$$

$$\cos \tfrac{1}{2} C = \sqrt{\frac{\sin s \sin (s-c)}{\sin a \sin b}}. \tag{99}$$

$$\tan \tfrac{1}{2} A = \sqrt{\frac{\sin (s-b) \sin (s-c)}{\sin s \sin (s-a)}}. \tag{100}$$

$$\tan \tfrac{1}{2} B = \sqrt{\frac{\sin (s-c) \sin (s-a)}{\sin s \sin (s-b)}}. \tag{101}$$

$$\tan \tfrac{1}{2} C = \sqrt{\frac{\sin (s-a) \sin (s-b)}{\sin s \sin (s-c)}}. \tag{102}$$

§ 155. $$\sin \tfrac{1}{2} a = \sqrt{-\frac{\cos S \cos (S-A)}{\sin B \sin C}}. \tag{103}$$

$$\sin \tfrac{1}{2} b = \sqrt{-\frac{\cos S \cos (S-B)}{\sin C \sin A}}. \tag{104}$$

$$\sin \tfrac{1}{2} c = \sqrt{-\frac{\cos S \cos (S-C)}{\sin A \sin B}}. \tag{105}$$

$$\cos \tfrac{1}{2} a = \sqrt{\frac{\cos (S-B) \cos (S-C)}{\sin B \sin C}}. \tag{106}$$

$$\cos\tfrac{1}{2}b=\sqrt{\frac{\cos(S-C)\cos(S-A)}{\sin C\sin A}}. \qquad (107)$$

$$\cos\tfrac{1}{2}c=\sqrt{\frac{\cos(S-A)\cos(S-B)}{\sin A\sin B}}. \qquad (108)$$

$$\tan\tfrac{1}{2}a=\sqrt{-\frac{\cos S\cos(S-A)}{\cos(S-B)\cos(S-C)}}. \qquad (109)$$

$$\tan\tfrac{1}{2}b=\sqrt{-\frac{\cos S\cos(S-B)}{\cos(S-C)\cos(S-A)}}. \qquad (110)$$

$$\tan\tfrac{1}{2}c=\sqrt{-\frac{\cos S\cos(S-C)}{\cos(S-A)\cos(S-B)}}. \qquad (111)$$

§ 156. $$\frac{\sin\frac{1}{2}(A+B)}{\sin\frac{1}{2}(A-B)}=\frac{\tan\frac{1}{2}c}{\tan\frac{1}{2}(a-b)}. \qquad (112)$$

§ 157. $$\frac{\cos\frac{1}{2}(A+B)}{\cos\frac{1}{2}(A-B)}=\frac{\tan\frac{1}{2}c}{\tan\frac{1}{2}(a+b)}. \qquad (113)$$

§ 158. $$\frac{\sin\frac{1}{2}(a+b)}{\sin\frac{1}{2}(a-b)}=\frac{\cot\frac{1}{2}C}{\tan\frac{1}{2}(A-B)}. \qquad (114)$$

$$\frac{\cos\frac{1}{2}(a+b)}{\cos\frac{1}{2}(a-b)}=\frac{\cot\frac{1}{2}C}{\tan\frac{1}{2}(A+B)}. \qquad (115)$$

ANSWERS.

§ 56; page 30.

12. 85° 56′ 37.32″.	**14.** 95° 29′ 34.8″.	**16.** 130° 55′ 5.952″.
13. 14° 19′ 26.22″.	**15.** 20° 27′ 2.52″.	

§ 63; page 39.

3. $n\pi,\ 2n\pi \pm \frac{\pi}{3}$.	**7.** $n\pi \pm \frac{\pi}{3}$.
4. $(2n+1)\frac{\pi}{2},\ n\pi + (-1)^n \frac{7\pi}{6}$.	**8.** $n\pi \pm \frac{\pi}{6}$.
5. $(2n+1)\frac{\pi}{2},\ n\pi \pm \frac{\pi}{4}$.	**9.** $n\pi,\ \pm \tan^{-1}\left(\frac{1}{7}\sqrt{7}\right)$.
6. $n\pi,\ n\pi \pm \frac{\pi}{4}$.	**10.** $\sin^{-1}\frac{\sqrt{5}-1}{2}$.

§ 76; page 44.

2. 1.5441.	**6.** 2.1003.	**10.** 2.5104.	**14.** 3.4192.
3. 1.6990.	**7.** 2.2922.	**11.** 2.5774.	**15.** 3.7814.
4. 1.6232.	**8.** 2.3892.	**12.** 2.9421.	**16.** 4.0794.
5. 1.8751.	**9.** 2.3222.	**13.** 2.8363.	**17.** 4.2006.

§ 78; page 44.

2. 0.5229.	**5.** 1.1549.	**8.** 0.2831.	**11.** 1.4592.
3. 0.2431.	**6.** 0.2589.	**9.** 0.7939.	**12.** 1.3468.
4. 1.6532.	**7.** 2.3522.	**10.** 2.1303.	**13.** 2.0424.

§ 81; page 45.

3. 3.3397.	**8.** 0.5663.	**13.** 0.6171.	**19.** 0.8752.
4. 1.7475.	**9.** 0.0430.	**14.** 0.2918.	**20.** 0.0794.
5. 0.6338.	**10.** 0.1165.	**16.** 0.0495.	**21.** 0.4248.
6. 8.6826.	**11.** 0.0939.	**17.** 0.0365.	**22.** 0.1051.
7. 1.0460.	**12.** 0.5440.	**18.** 0.7007.	**23.** 0.0406.

§ 85; page 47.

2. 0.5562.
3. 1.0491.
4. 9.9242 – 10.
5. 8.9912 – 10.
6. 7.5353 – 10.
7. 3.4592.
8. 8.5932 – 10.
9. 6.6074 – 10.
10. 9.2885 – 10.
11. 2.3064.
12. 0.1151.
13. 0.7782.

§ 86; page 47.

4. 0.011739.
5. 2.527511.
6. 6.780210 – 10.
7. 4.812917.
8. 3.960116.
9. 7.013152 – 10.
10. 4.942550 – 10.
11. 5.863566.
12. 5.640409 – 10.
15. 6.61005.
16. 55606.5.
17. .0110890.
18. 186.334.
19. .00223905.
20. .0000100006.
21. 9776.67.
22. 467929.
23. .000342770.
24. .00000130514.

§ 91; pages 50, 51.

1. 1897.85.
2. – 193315.
3. .309170.
4. .00110375.
5. 6.36103.
6. .0301742.
7. 31.2004.
8. – .132693.
9. .126965.
10. .0235770.
11. – 1.16493.
12. – .00256105.
13. 3692.77.
14. .277996.
15. – 15896.0.
16. .0316228.
17. 244.004.
18. .00279116.
19. .000000237177.
20. 2.23607.
21. 1.14870.
22. – 1.22028.
23. 1.77828.
24. .668289.
25. .645831.
26. .137751.
27. – .370134.
30. 13.8289.
31. 2.48722.
32. 1.05557.
33. .0000214279.
34. .00710469.
35. .695490.
36. .542699.
37. – 36.0189.
38. – 11.1122.
39. .943241.
40. 2.62762.
41. 2.53217.
42. – 1.79616.
43. 1.03242.
44. .298557.
45. .0448607.
46. .794509.
47. 1.80492.
48. 179.596.
49. 1.88270.
50. .000193152.
51. – .0995935.
52. 1.34384.

§ 92; page 52.

3. $x = .2831+$.
4. $x = -2.173+$.
5. $x = 1.155+$.
6. $x = -.1765+$.
7. $x = \frac{5 \log c}{\log a - 2 \log b}$.
8. $x = \frac{3 \log a}{4 \log n - 2 \log m}$.
9. $x = \frac{1}{2}$.
10. $x = 1$ or -5.

§ 93; page 52.

2. 3.7004+. **3.** −.06546+. **4.** −6.059+. **5.** 3.326+.

6. −.4601+. **7.** .3494+. **9.** 4. **10.** $\frac{5}{3}$. **11.** $-\frac{1}{3}$. **12.** $\frac{6}{5}$.

§ 94; page 53.

1. 9.345950 − 10.
2. 0.376890.
3. 9.932630 − 10.
4. 9.865995 − 10.
5. 9.243533 − 10.
6. 9.163433 − 10.
7. 0.302190.
8. 0.153906.
9. 0.002256.
10. 59° 15′ 26.4″.
11. 33° 0′ 16.1″.
12. 81° 7′ 37.9″.
13. 27° 31′ 50.5″.
14. 8° 41′ 32.7″.
15. 75° 45′ 9.8″.
16. 49° 38′ 57.1″.
17. 23° 26′ 30.9″.

§ 95; page 53.

1. .68573.
2. .25232.
3. .06344.
4. .69518.
5. .92163.
6. .86962.
7. 51° 36′ 42.9″.
8. 15° 28′ 22.5″.
9. 66° 14′ 34.3″.
10. 29° 9′ 13.8″.

§ 96; page 53.

1. 8.338076 − 10.
2. 8.810945 − 10.
3. 1.369926.
4. 0° 58′ 51.06″.
5. 0° 24′ 53.79″.
6. 1° 37′ 41.93″.

§ 102; pages 56 to 58.

1. $a =$ 1.8117, $b =$ 6.7615.
2. $b =$ 11.7793, $c =$ 12.7965.
3. $a =$ 16.7820, $c =$ 26.1081.
4. $A =$ 34° 22′ 7.1″, $b =$ .511764.
5. $A =$ 33° 8′ 56.3″, $c =$ 499.252.
6. $b =$ 10.3547, $c =$ 13.1404.
7. $a =$ .0036235, $b =$ .013523.
8. $A =$ 39° 49′ 24.6″, $a =$ 48.8645.
9. $a =$ 148.407, $c =$ 948.680.
10. $A =$ 49° 53′ 54.9″, $c =$ 4.46330.
11. $b =$ .000336374, $c =$ .00336715.
12. $a =$ 3821.55, $b =$ 3641.34.
13. $A =$ 35° 53′ 55.2″, $b =$ 731.237.
14. $a =$ 176.533, $c =$ 191.993.
15. $a =$ 20455.6, $c =$ 21405.6.
16. $a =$ 2.40989, $b =$ .812578.
17. $A =$ 19° 31′ 57.2″, $c =$ .000505172.
18. $b =$ 77.6330, $c =$ 91.2952.
19. $A =$ 32° 10′ 16.5″, $a =$ 388.471.
20. $b =$ 644.109, $c =$ 650.272.
21. $a =$ 34308.0, $b =$ 23381.6.
22. $b =$ 4.48174, $c =$ 8.5085.
23. $A =$ 39° 21′ 54.1″, $b =$ 121.240.
24. $a =$ .00247181, $c =$ .00360016.
25. $a =$ 16001.6, $c =$ 85725.1.
26. $a =$ 3624500, $b =$ 8821960.

27. $A = 76° 33' 49.0''$, $a = 24234.4$.
28. $a = 207302$, $b = 421170$.
29. $a = .507624$, $c = .525355$.
30. $A = 60° 14' 12.9''$, $c = 774.563$.
31. $c = 252.103$.
32. $a = 1.73561$.
33. $c = 122748$.
34. $A = 47° 42' 47.8''$.
35. $a = .344647$.
36. $a = 4925.31$.
37. 20.573.
38. 83.271 ft.
39. 31° 47′ 24.5″.
40. 36° 37′ 58.0″.
41. 99.4565 mi.
42. 10.2352.
43. 19° 49′ 46.7″.
44. 365.64 ft.
45. 56° 18′ 35.7″.
46. 25.2230 mi., 30.0750 mi.
47. 21.6514.
48. 14.4853, 15.6787.
49. 517.51 ft.
50. 17.2624.
51. 420.867 ft.
52. 437.605.
53. 10.392.
54. 482.1 ft.
55. Rate, 6.79668 miles an hour; bearing, N. 63° 8′ 28.5″ W.

§ 104; page 60.

2. $B = 89° 59' 42.8''$.
3. $B = 89° 23' 22.6''$.
4. $A = 89° 59' 37.2''$.
5. $B = 89° 59' 59.0''$.
6. $A = 89° 43' 13.6''$.

§ 106; page 61.

2. 6.9066.
3. .151079.
4. 5699.7.
5. .089433.
6. 8130.9.
7. .0067825.
8. 2.18876.
9. 107.762.
10. .0487840.

§ 114; page 67.

2. $b = 283.331$, $c = 267.677$.
3. $a = .340132$, $c = .986084$.
4. $a = 29.0595$, $b = 18.3742$.
5. $a = .0313440$, $c = .0498733$.
6. $b = 5.76721$, $c = 2.16917$.
7. $a = 5058.5$, $b = 3683.53$.
8. $a = .299674$, $b = .731538$.
9. $a = 4.01036$, $c = 3.55195$.
10. $b = 56719.9$, $c = 23073.5$.

§ 115; pages 68, 69.

2. $A = 118° 17' 57.4''$, $b = 44.7274$.
3. $A = 60° 44' 39.5''$, $c = 965.282$.
4. $C = 63° 49' 9.3''$, $a = 4.48237$.
5. $B = 28° 43' 49.0''$, $c = 1.44246$.
6. $B = 145° 35' 24.7''$, $a = 1045.74$.
7. $C = 63° 48' 28.1''$, $b = 13.7387$.
8. $A = 67° 55' 16.9''$, $c = 85.3596$.
9. $C = 46° 13' 20.9''$, $a = .0759588$
10. $C = 134° 36' 27.4''$, $b = 27335.0$.

§ 116; page 70.

3. $A = 28° 57' 18.0''$, $B = 46° 34' 2.8''$, $C = 104° 28' 39.0''$.

4. $A = 44° 24' 54.8''$, $B = 78° 27' 47.0''$, $C = 57° 7' 17.6''$.

5. $A = 71° 47' 24.4''$, $B = 58° 45' 5.4''$, $C = 49° 27' 30.0''$.

6. $A = 74° 40' 16.4''$, $B = 47° 46' 39.0''$, $C = 57° 33' 4.8''$.

7. $A = 59° 19' 11.8''$, $B = 68° 34' 7.6''$, $C = 52° 6' 40.6''$.

8. $A = 45° 11' 46.6''$, $B = 101° 22' 17.8''$, $C = 33° 25' 56.4''$.

9. $A = 71° 33' 49.2''$. 10. $B = 30° 47' 22.8''$. 11. $C = 25° 56' 54.2''$.

§ 121; pages 73, 74.

1. $B = 32° 36' 9.4''$, $c = 6.62085$.
2. $B_1 = 31° 57' 47.8''$, $a_1 = 120.313$;
   $B_2 = 148° 2' 12.2''$, $a_2 = 11.3800$.
3. $C = 23° 33' 18.2''$, $a = .183882$.
4. $A = 34° 29' 48.2''$, $b = 7.12905$.
5. Impossible.
6. Impossible.
7. $B = 48° 34' 38.4''$, $a = 76.0172$.
8. $C = 90°$, $b = 5.51109$.
9. $C_1 = 46° 18' 35.5''$, $a_1 = 6.94575$;
   $C_2 = 133° 41' 24.5''$, $a_2 = .699906$.
10. $A = 25° 32' 50.9''$, $c = 278.193$.
11. Impossible.
12. $C = 14° 4' 7.7''$, $b = 1.43516$.
13. $B = 90°$, $c = 137.872$.
14. $A_1 = 70° 12' 46.7''$, $b_1 = .287904$;
    $A_2 = 109° 47' 13.3''$, $b_2 = .104539$.
15. $B = 45° 38' 30.2''$, $a = 16214.3$.

§ 122; pages 74, 75.

2. 197.656. 5. 165917. 8. .078614. 11. 4000.81.
3. 14.9812. 6. 2878.31. 9. 860.006. 12. .000329015.
4. 16.6843. 7. 1.30108. 10. .0448746. 13. 25.6249.

§ 123; pages 75, 76.

1. Height, 153.629 ft.; distances, 117.246 ft., 217.246 ft.

2. $AD = 44.9525$. 4. $47° 52' 2.1''$. 6. 56.6547, 49.3482.

3. 29799.9 sq. rd. 5. 247.741 ft. 7. 32.5255 mi.

8. Two angles, 74° 12′ 20.0″, 58° 23′ 48.0″; third side, .430133.

9. N. 47° 32′ 33.1″ W.

10. 9.8995 mi., 19.1244 mi.

11. One angle, 101° 13′ 45.8″; diagonal, 136.187.

12. 297.954 ft.

13. Sides, 26.5604, 90.5154; one angle, 119° 5′ 14.6″.

14. 91.6364 ft., 33.8973 ft.

15. 17.64934, 8.77461.

16. 1113.34 ft.

17. Diagonal, 52.9024; side, 41.9506.

18. 247.998 ft.

19. $AD = 88.1534$, $A = 56° 1′ 10.7″$.

20. 1569.948 sq. rd.

§ 126; page 79.

2. 2.11491, −1.86081, −.254102.

3. 2.14510, .523978, − 2.66907.

4. .47761, − 6.1364, − .34120.

5. 3.49086, − .83425, .343379.

§ 148; page 93.

5.	$A =$ 36° 58′ 50.0″,	$B =$ 63° 42′ 34.0″,	$b =$ 42° 34′ 54.4″.
6.	$a =$ 27° 49′ 17.9″,	$b =$ 42° 29′ 21.8″,	$c =$ 49° 17′ 42.4″.
7.	$B =$ 68° 37′ 18.1″,	$b =$ 44° 56′ 46.7″,	$c =$ 49° 20′ 41.8″;
or,	$B =$ 111° 22′ 41.9″,	$b =$ 135° 3′ 13.3″,	$c =$ 130° 39′ 18.2″.
8.	$A =$ 68° 10′ 4.4″,	$b =$ 163° 42′ 32.1″,	$c =$ 141° 50′ 15.2″.
9.	$A =$ 15° 34′ 32.3″,	$B =$ 94° 14′ 40.0″,	$c =$ 105° 26′ 27.5″.
10.	$a =$ 170° 13′ 25.6″,	$B =$ 78° 34′ 3.4″,	$b =$ 40° 1′ 8.6″.
11.	$A =$ 21° 11′ 12.7″,	$a =$ 19° 50′ 30.4″,	$c =$ 69° 54′ 41.6″;
or,	$A =$ 158° 48′ 47.3″,	$a =$ 160° 9′ 29.6″,	$c =$ 110° 5′ 18.4″.
12.	$A =$ 82° 8′ 19.3″,	$a =$ 73° 38′ 54.4″,	$b =$ 28° 4′ 23.5″.
13.	$A =$ 122° 34′ 33.5″,	$a =$ 132° 24′ 39.6″,	$B =$ 52° 58′ 9.5″.
14.	$A =$ 153° 10′ 2.8″,	$B =$ 115° 25′ 2.8″,	$c =$ 20° 2′ 40.3″.
15.	$A =$ 165° 50′ 26.0″,	$b =$ 139° 10′ 11.5″,	$c =$ 41° 42′ 23.4″.
16.	$a =$ 112° 16′ 49.7″,	$b =$ 145° 51′ 35.5″,	$c =$ 71° 42′ 41.1″.
17.	$A =$ 55° 58′ 5.5″,	$B =$ 34° 41′ 20.4″,	$c =$ 12° 39′ 44.7″.
18.	$a =$ 54° 0′ 24.8″,	$B =$ 84° 43′ 10.5″,	$c =$ 86° 10′ 32.3″.
19.	$a =$ 41° 29′ 25.7″,	$b =$ 133° 39′ 29.8″,	$c =$ 121° 8′ 21.5″.
20.	$a =$ 152° 35′ 19.0″,	$B =$ 108° 7′ 8.6″,	$b =$ 125° 24′ 13.7″.
21.	$A =$ 20° 3′ 21.5″,	$a =$ 14° 58′ 21.1″,	$c =$ 131° 7′ 4.9″;
or,	$A =$ 159° 56′ 38.5″,	$a =$ 165° 1′ 38.9″,	$c =$ 48° 52′ 55.1″.
22.	$a =$ 110° 57′ 15.6″,	$B =$ 165° 10′ 31.9″,	$c =$ 69° 41′ 7.1″.
23.	$A =$ 111° 53′ 21.2″,	$B =$ 115° 40′ 6.8″,	$b =$ 117° 49′ 41.2″.

24.	$A=165°\ 3'\,57.9''$,	$a=168°\ 8'\,48.3''$,	$b=51°\,53'\,53.3''$.
25.	$B=22°\,13'\ 3.9''$,	$b=20°\,34'\,38.3''$,	$c=111°\,38'\,31.1''$;
or,	$B=157°\,46'\,56.1''$,	$b=159°\,25'\,21.7''$,	$c=68°\,21'\,28.9''$.
26.	$A=64°\,30'\,52.0''$,	$a=38°\,32'\,30.5''$,	$B=146°\,37'\,27.3''$.

§ 149 ; page 94.

2.	$a=103°\,25'\,57.4''$,	$B=157°\,31'\,44.4''$,	$C=119°\,19'\,11.3''$.
3.	$a=57°\,43'\,57.2''$,	$b=129°\,56'\,31.7''$,	$C=58°\ 4'\,55.6''$.
4.	$A=19°\,56'\,45.0''$,	$B=141°\,38'\,20.3''$,	$b=113°\,18'\,58.3''$.
5.	$A=44°\,41'\,15.9''$,	$a=51°\,37'\ 1.9''$,	$b=60°\,51'\ 3.4''$.
6.	$B=80°\,27'\,25.7''$,	$b=80°\,46'\,54.3''$,	$C=87°\,31'\,12.5''$;
or,	$B=99°\,32'\,34.3''$,	$b=99°\,13'\ 5.7''$,	$C=92°\,28'\,47.5''$.
7.	$A=67°\,11'\,45.0''$,	$B=80°\,58'\,16.5''$,	$C=93°\,29'\,13.4''$.

§ 150 ; page 95.

2. $a=69°\,55'\,43.2''$, $C=159°\,59'\,40.6''$.
3. $A=120°\,41'\,19.6''$, $c=30°\,14'\,37.4''$.
4. $A=140°\,35'\ 4.5''$, $C=145°\,11'\,50.4''$.
5. $C=148°\,19'\,24.8''$, $c=80°\,47'\,39.8''$.

§ 161 ; page 104.

2.	$a=95°\,37'\,51.0''$,	$b=41°\,52'\,22.2''$,	$C=110°\,48'\,24.0''$.
3.	$b=98°\,30'\,32.4''$,	$c=56°\,42'\,47.0''$,	$A=59°\,38'\,53.2''$.
4.	$c=64°\,19'\,27.8''$,	$a=34°\ 3'\,11.8''$,	$B=37°\,39'\,27.2''$.
5.	$b=146°\,25'\ 1.4''$,	$a=69°\ 4'\,38.2''$,	$C=125°\,11'\,41.8''$.

§ 162 ; page 105.

2.	$A=121°\,32'\,41.3''$,	$B=40°\,56'\,48.5''$,	$c=37°\,25'\,48.8''$.
3.	$A=86°\,59'\,48.8''$,	$C=60°\,50'\,54.8''$,	$b=111°\,16'\,42.4''$.
4.	$C=134°\,57'\,31.3''$,	$B=50°\,40'\,48.3''$,	$a=69°\ 7'\,34.6''$.
5.	$B=163°\ 8'\,48.4''$,	$A=147°\,29'\,24.2''$,	$c=76°\ 8'\,49.0''$.

§ 163 ; page 106.

2.	$A=51°\,58'\,28.0''$,	$B=58°\,53'\,13.2''$,	$C=83°\,54'\,31.6''$.
3.	$A=142°\,32'\,37.8''$,	$B=27°\,52'\,36.0''$,	$C=32°\,26'\,52.8''$.
4.	$A=142°\,23'\,44.0''$,	$B=159°\,15'\,41.6''$,	$C=133°\,14'\ 4.2''$.
5.	$A=47°\,21'\,11.8''$.		

§ 164; page 107.

3. $a = 68° 46' 28.4''$, $b = 73° 47' 57.8''$, $c = 63° 12' 24.6''$.

4. $a = 90° 53' 2.6''$, $b = 117° 48' 59.6''$, $c = 132° 5' 10.0''$.

5. $a = 103° 31' 33.8''$, $b = 53° 4' 26.2''$, $c = 61° 14' 18.2''$.

6. $b = 85° 48' 53.8''$.

§ 165; page 109.

4. $C = 65° 29' 1.0''$, $A = 97° 18' 33.8''$, $a = 100° 42' 23.4''$.

5. $B = 42° 40' 9.2''$, $C = 159° 54' 3.6''$, $c = 153° 29' 39.8''$;
or, $B = 137° 19' 50.8''$, $C = 50° 21' 16.4''$, $c = 90° 8' 51.4''$.

6. Impossible.

7. $C = 90°$, $B = 113° 33' 15.5''$, $b = 114° 47' 47.5''$.

8. $B = 68° 17' 2.4''$, $A = 132° 35' 12.4''$, $a = 131° 16' 32.2''$;
or, $B = 111° 42' 57.6''$, $A = 77° 3' 48.0''$, $a = 95° 48' 41.8''$.

9. Impossible.

10. $C = 146° 37' 40.2''$, $B = 55° 1' 11.8''$, $b = 96° 33' 16.2''$.

§ 166; page 111.

2. $b = 114° 48' 57.9''$, $a = 82° 54' 0.0''$, $A = 79° 18' 29.0''$.

3. $a = 67° 25' 2.3''$, $c = 160° 6' 10.0''$, $C = 164° 6' 8.4''$;
or, $a = 112° 34' 57.7''$, $c = 103° 6' 20.4''$, $C = 128° 22' 54.8''$.

4. $c = 90°$, $B = 63° 46' 30.2''$, $b = 66° 29' 37.6''$.

5. Impossible.

6. $b = 27° 22' 7.6''$, $a = 117° 9' 5.2''$, $A = 47° 20' 57.2''$.

7. $a = 43° 2' 23.6''$, $b = 129° 9' 46.0''$, $B = 89° 23' 51.8''$;
or, $a = 136° 57' 36.4''$, $b = 20° 34' 54.2''$, $B = 26° 57' 36.4''$.

8. Impossible.

§ 167; page 112.

1. Distance, 3275.20 mi.; bearing of Boston from Greenwich, N. 71° 38′ 53.7″ W.; of Greenwich from Boston, N. 53° 6′ 31.9″ E.

2. Distance, 11012.9 mi.; bearing of Calcutta from Valparaiso, S. 64° 20′ 17.4″ E.; of Valparaiso from Calcutta, S. 54° 54′ 25.2″ W.

3. Latitude, 49° 58′ 23.1″ N.

§ 170; page 113.

1. Time, 6 h. 0 m. 43 s. A.M.; longitude, 44° 49′ 18″ W.

2. 15° 0′ 41.4″. 3. N. 56° 28′ 8.5″ E. 4. 5 h. 3 m. 27 s. A.M.